D1208231

FOR **FOOD, PROFIT,** AND **FUN**

YVONNE ZWEEDE-TUCKER

First published in 2011 by Voyageur Press, an imprint of MBI Publishing Company, 400 First Avenue North, Suite 300, Minneapolis, MN 55401 USA

All photographs are from the author's collection unless noted otherwise.

Library of Congress Cataloging-in-Publication Data

Zweede-Tucker, Yvonne, 1961–
The meat goat handbook : raising goats for food, profit, and fun / Yvonne Zweede-Tucker. — 1st ed.
p. cm.
Includes index.
ISBN 978-0-7603-4042-4 (flexibound)
1. Goats—Handbooks, manuals, etc. 2. Meat animals—Handbooks, manuals, etc. I. Title.
SF383.Z94 2012
636.3'913—dc23

2011017215

Editors: Danielle Ibister, Melinda Keefe, and Anitra Budd
Design Manager: LeAnn Kuhlmann
Layout: Mandy Kimlinger
Designer: Cindy Samargia Laun
Cover designer: Matthew Simmons
Goat illustrations: Gary DeWitt, except where noted otherwise

On the frontis:
Staci McStotts, Critics Choice Pygmy Goats

On the title page:
Competing in shows with meat goats can offer young people valuable experience for future social or business endeavors. *Richard and Sandy's Boer Goat Farm*

Printed in China

10 9 8 7 6 5 4 3 2 1

CONTENTS

INTRODUCTION

WHY MEAT GOATS?

Welcome to the wonderful world of meat goats! You're probably attracted to the idea of raising meat goats for a variety of reasons. Perhaps you're intrigued by their personalities—their intelligence and spontaneity makes them fun animals to raise. They almost seem to have a sense of humor, and as long as their basic needs are met, they enjoy life fully.

Maybe you're interested in raising meat goats because they're profitable. The United States and Canada don't produce as much goat meat as consumers wish to purchase. Due to the scarcity of supply, the value of the meat continues to rise, making its production easier to carry out profitably. Many different cuisines use goat meat either as a staple for everyday meals or as a delicacy for holidays or special occasions. In fact, approximately 80 percent of the world's population consumes goat meat regularly, according to the University of California Cooperative Extension. Not only

The goat in world history

A magnificent ancient Byzantine mosaic of a shepherd milking a goat and another farmer coming to collect it in a large jar. From the remains of the great palace in Constantinople (modern-day Istanbul, Turkey). *Shutterstock*

is the number of people who already prefer goat meat growing, but every day new consumers are discovering the delicious flavor and health benefits of ***chevon*** or ***cabrito*** (other names for goat meat).

You might be attracted to meat goats' manageability. Since they're relatively small animals, they can be raised on a reasonable scale (several goats on one or a few acres of land). Meat goats can't accidentally hurt you as easily as large animals can, and they can be well managed by a person working alone. Hardy and largely undemanding animals, meat goats are not only perfect homestead livestock, but can also be successfully managed in large numbers.

It could be that you're simply seeking a change of pace. Meat goats can be a rewarding first career, as well as a wonderful new career after retirement if you long to be outside and with animals. Opportunities abound in the meat goat industry, not only in animal production but in all facets of marketing, transportation, genetics, and management.

Whatever your reasons for wanting to raise meat goats, you'll find that they provide joy and comfort to all who own them. But don't just take it from me. Here's what other producers had to say when I asked them why they started (and kept going) with meat goats:

- "I want to work outside with my daughters, learning together and creating something that our customers will be pleased to acquire or consume."
- "I hadn't realized how much we appreciate the intelligence and personality of goats. It makes raising them so much more fun than business just for the profit of it. So entertaining!"
- "Goats help with each of the things that I want to do in my life: improve the soil every year, have plenty of delicious and nutrient-dense food, be surrounded by cheerful beings, learn until the last moment of my life, and have enough money that I don't have to worry."
- "We do it because it makes our souls sing and our life worthwhile. We are not complete without the rhythm of life and the song of the seasons."

As you begin your journey with meat goats, you'll soon find that there is a lot that you want (and need) to learn about them. That's where *The Meat Goat Handbook* comes in.

THE MIND OF A GOAT

The domestic goat, which includes our beloved meat goats, has an extremely adaptable, clever, entertaining, fascinating (and occasionally exasperating!) mind. Adaptability is defined as being able to change to take advantage of the environment. When you consider all of the different farms and ranches across North America where meat goats are surviving and thriving, adaptability isn't just a description of the trait that helps them survive—it's a good description of how amazing these animals are. Some meat goats live happily in areas with very little annual rainfall and temperatures that frequently rise well above 100 degrees Fahrenheit in the summer. Other meat goats live happily in areas with heavy annual rainfall and humidity, or in areas where over half the year is considered winter and temperatures plummet far below zero. I think that not only could goats well have coined the phrase "just deal with it," but also that they deal with life with an admirable amount of verve and spirit.

ABOUT THIS BOOK

If you're curious about, thinking about, or planning to start a meat goat enterprise, you probably have lots of questions: What kind of goats should you choose? How do you buy them and care for them? How do they reproduce? How do you sell them?

Having faced all of those questions and more, I've learned firsthand that there's usually more than one right answer. In *The Meat Goat Handbook*, I'll tell you what has worked for me and my husband and for other producers. I'll show you how to clarify your goals, whether they be chuckling many times per day or paying off your land. I'll explain the ins and outs of preparing for, acquiring, and caring for your goats, allowing you to hit the ground running rather than scrambling to catch up. I'll also offer practical tips on a variety of related topics, from protecting your goats from predators to breeding and raising kids to understanding and managing your enterprise, even if it's for pleasure. *The Meat Goat Handbook* isn't breed specific; it's success specific!

In the Resources section at the end of this book, I've included a list of useful periodicals, books, and websites to help you get started in your meat goat enterprise. Following the Resources section is the Glossary, one of my favorite parts of the book (after all the beautiful pictures graciously shared by so many in the industry). In it you'll find definitions for the most common words used in the meat goat industry. The first time a Glossary word is used in the book, it will be bolded (see, you were probably wondering why a word was bolded earlier). There also are handy management tips and information incorporated into the Glossary.

Why do I use the word "enterprise" rather than "job" or "business" when I refer to meat goats? Because I don't think you should call something that's this fun a job, even if you're earning money from it. Because for most of us, the word "business" has many deep-rooted overtones, of which humor isn't one. And because the dictionary says that enterprise is "readiness to embark on bold new ventures" and "a purposeful or industrious undertaking, especially one that requires effort or boldness." I believe that if you're bold and ready to take on a new challenge with meat goats, *The Meat Goat Handbook* will be an invaluable guide.

Please enjoy *The Meat Goat Handbook*!

The Meat Goat Handbook—one of the best illustrated, well-written, and easy-to-read books on meat goat production—comes from a highly experienced "goat lady" who has not only been raising and breeding goats for 20 years but mentoring and helping new goat producers as well. The book has a wealth of management and economic information as well as outside-the-box, challenging ideas that will be valuable to more-experienced producers. It is the only goat book written from the standpoint of having a profitable commercial meat goat enterprise. If you are a goat producer—whether new, wannabe, or experienced—this is the best book for learning about meat goat production.

Dr. Steve Hart
Goat Extension Specialist
Langston, Oklahoma

The goat in history right here in America
Petroglyphs of what could be goats adorn the canyon walls in Utah's Capitol Reef National Park, chipped into the stone over a thousand years ago by the Fremont people. *Petroglyphtrail*

CHAPTER 1

ESTABLISHING GOALS

What do you want out of your meat goat enterprise? You probably want to be successful, satisfied, and proud of what you do in the world of meat goats.

What is success? You get to decide what success is for you: whether you want to make money with your goats, have fun, save your corner of the world from invasive plants, protect a rare or endangered breed, help your children learn how to care for livestock, produce healthy and delicious meat, or a combination of all of those. Whatever your goal, you get to decide what you want to do and how.

How do you set yourself up for success? I suggest deciding on your main goal in raising meat goats, then decide if there is anything else that you really want to get out of your goat enterprise. Be honest with yourself and the people who are important to you about what is most important, then what's second, then what's third. Perhaps you want your goat enterprise to break even or your goat kids to grow to a certain size by a certain age (measured in months or even days). Or perhaps your goal is to have your goats raise the most salable pounds of market kids per pound of **doe** per acre of land per year. You'll soon find that if you make dramatic progress toward one goal, there is usually a cost in progress toward another goal. Taking short-term profit (for example, by selling your does or doe kids as market goats for

The Boer goat has become the standard in market meat goats. Producers can select for the attributes that make a goat perfect for their environment and management system. *Lynn Stone*

No matter what your financial goal is, knowing how your revenues and costs compare to each other, and why, is extremely important to overall satisfaction with your goat enterprise.

When one goat comes running to greet you, usually they all do. Herd instinct makes meat goats want to stay together. *Bryce Gromley for Gromley Farms*

meat) will make it impossible for you to have kids born from those goats in the future. Also, doe kids raised at faster rates of weight gain typically don't have the same fertility as goats with a more moderate rate of growth. That's why it's important to prioritize your goals and be realistic about achieving one at a time.

No matter what you want to achieve, stay focused on that, and judge every opportunity to spend or make money on whether it will get you closer to your goal(s). Pretend that you have a yardstick to hold up against things and see if they measure up so that you can move closer to your target. It's better in the long run to pass up an opportunity than to regret having gotten overwhelmed or stuck.

So what are some goals that you can achieve with meat goats?

Meat goats are more than happy to consume unwanted vegetation that few other animals will eat, such as these honey locust trees. *Jean-Marie Luginbuhl, North Carolina State University*

ECONOMIC PROFIT

You goal may be (1) to have money left over after you pay all your goat bills, (2) to break even by spending as much money as your goats bring in, or (3) to lose the least amount of money possible with your goat enterprise. No matter what your financial goal is, knowing how your revenues and costs compare to each other, and why, is extremely important to overall satisfaction with your goat enterprise.

There are many opportunities for people in the meat goat industry. Raising market goats, raising goats for breeding stock, having goats do forage-management projects, working as an organizer to match goats from different producers to the needs of different customers, or designing websites or advertising programs are all ways that you can match your unique talents to the needs of the growing world of meat goats. The easiest way to have money left at the end of each year is to spend just a few minutes planning before you buy your first goats, and checking on your progress a few times per year. A startup financial worksheet is outlined in Chapter 11.

ENVIRONMENTAL BENEFIT

Goats have a wonderful advantage over many other domesticated forage-consuming animals in that they will search out, and devour with gusto and glee, plants that are usually thought of as a nuisance. Goats adore (and most farmers and ranchers have no love for) wild roses, kudzu, thistles, poison ivy and oak, leafy spurge, spotted knapweed, shrubby cinquefoil, yellow star thistle, buttercups, oxeye daisies, houndstongue, osage orange, and honey locust.

Goats have a wonderful advantage over many other domesticated forage-consuming animals in that they will search out, and devour with gusto and glee, plants that are usually thought of as a nuisance.

Meat goats will consume these undesired plants and turn them into a red meat that is in short supply and in growing demand. Some

MEAT GOATS AND STEWARDSHIP

Tom and Meta Syfan of Mountain Home, Texas, have been in the meat goat business since 1955. When complimented on how beautiful their ranch and animals look (the Syfans raise black Spanish goats, fine-wool sheep, and crossbred cattle), Tom smiled and said, "We only get to borrow the land and the animals while we're here. Stewardship is what we do with them while we have them borrowed." After two decades of raising meat goats, up until that moment I thought that I was getting somewhat knowledgeable about the goats. Tom's quiet statement made me see clearly that there was still so much to learn.

Meat goat producers can raise goat breeds such as this Savanna doe and her new kid. Savannas are one of the newest breeds to become available in the United States. *Priscilla Ireys, Critton Creek Farm*

Cashmere is one of the finest fibers available and is produced by a meat goat. Both cashmere and goat meat are in short supply in the United States. *Anne Marcom for Cozine Springs Ranch*

people even get paid to have their goats practice biological control of undesired species of plants. Others are carrying out land restoration by using goats to improve damaged land. Meat goats can assist land managers in achieving biodiversity—using different species of animals to harvest what nature produces rather than trying to make land produce what we want.

ENRICHING YOURSELF AND THE WORLD

You don't have to pick goals only for yourself when it comes to meat goats—the whole family can benefit from raising them. Meat goats are smart, funny, and affectionate and many will wag their tails when you pat them. Goats can give young members of a family livestock experience with animals that are a very manageable

Angora goats can be raised for the luxurious mohair fiber that they produce, in addition to being raised as meat animals. They must be shorn annually. *Karen Stieler*

size. Children can have goats for 4-H, learning about animal husbandry, record keeping, and showmanship.

Adults of all ages also can find satisfaction with meat goats. They can work on importing additional examples of a sought-after breed, protecting a rare or endangered breed, or producing more animals to feed consumers seeking goat meat for its delicious flavor and health benefits.

You can breed goats for a specific market, whether it be goats of a specific size, color pattern, fertility, longevity, or attitude (mild-mannered or tough enough to survive almost anything nature throws at them). Meat goats can produce cashmere, one of the world's finest and most luxurious fibers. Angora goats produce mohair, a lustrous specialty fiber.

Successful livestock management skills are not inherited, they are learned.

START SMART

Reaching your goals isn't difficult, if you know where you want to go and keep your attention on your progress. Setting up and operating a goat enterprise, whether you want to make money, have fun, or improve the world, can be a rewarding experience. Most advice will tell you to start small and let your herd size grow after you learn the ins

The Spanish landrace goat, once nearly lost to crossbreeding, is making a comeback due to the breed's hardy constitution and longevity. *Jeannette Beranger, American Livestock Breeds Conservancy*

Most goats love their people. Raising livestock is nice when you are appreciated and doing chores isn't a chore! *Liberty Pastures*

This doe seems to be asking, "How much longer until kidding season?" She delivered triplets a few days after this photo was taken and raised them herself without any assistance.

Market goats wait in pens to be sold at a major livestock auction in Texas. The pens have roofs to shade the animals from direct sun. *Carol DeLobbe, Bon Joli Farm*

and outs of raising and selling meat goats through "on the job" training. There is, in fact, no way to replace experience in learning how to do things right, but since my experiences with meat goats have almost all been good, I will be very honest with you and tell you what we would do differently if we were able to do it all over.

This book will help you to "start smart," whether you want a small, medium, or large herd of goats. Successful livestock management skills are not inherited, they are learned. In *The Meat Goat Handbook*, you'll read about the lessons that my husband and I have learned over many

Selling market kids is one possible source of income for your meat goat operation. Here, a small group of market kids is brought to a sale facility. *Carol DeLobbe, Bon Joli Farm*

A unique coat color pattern is attractive to some breeding stock buyers. Meat goats come in solid colors, two tone, "paints," and dapples, so there is a pattern for every customer's preference. *Carol DeLobbe, Bon Joli Farm*

Shrubby cinquefoil (*potentilla*) is a nuisance to cattle ranchers, as cows don't like it and thus it spreads across rangeland. However, meat goats will consume cinquefoil before they eat grass!

years, so that you know what to expect rather than getting any surprises on your journey. You already are well on your way to success: You have gotten this book and read this far and you're interested in or drawn to meat goats, which will make raising and selling goats fun. So keep up the good work!

ONE SIZE DOES NOT FIT ALL

There is no one right answer to the question of how many goats are right for your goat enterprise, what size goats are right for you, and what goals are right for you. So how do you decide what goals are right for you, choose the right goats to help you toward your goals, and determine how to evaluate your progress? It might seem boring, but moderation will get you to your destination. It's not as exciting and easy to brag about, but remember the story of the tortoise and the hare? Bankers and spouses are usually quite supportive of moderation too.

WHAT IS SUCCESS?

Success comes in many forms. Did one of your favorite goats finally have a female kid after previously only having males, so that soon you will be able to have that doe kid bred by a promising new **buck** that you just purchased? Has your county asked you to figure out how your goats could visit the hillside next to the city park and consume the Canada thistles that are growing there, as the hillside is too steep for human groundskeepers to tackle safely? Did a young person purchase a goat from you and then have fun showing it at the county fair? Because of the money that you have earned with the goats, do you have only a few years left on your land payment, or is the family farm once again profitable? Are you selling your market goats to an online store where people can buy cuts of *chevon* via the Internet? Any of these could be signs of success. It all depends on the goals you set for yourself and your goats.

Boer blood is evident in these meat goats, heading back out to pasture in Idaho after coming in to drink at their water trough. *Tangy and Matthew Bates, Blue Creek Livestock*

Goats and bushes are a great combination, according to the goat. Standing in knee-high grass, this young Spanish doe chooses to eat leaves rather than grass.

Meat goats to the rescue! This overgrown and unhealthy woodlot is being brought back to good health by a herd of hungry meat goats. Portable electric net fence concentrates the goats on the thick underbrush, and now they are mowing the grass near the trees. *Schneider Farms*

It almost looks like this goat kid is telling secrets to the child! Bottle baby goats offer unconditional love to their human foster parents. *Cori McKay for Smoke Ridge*

One thought that helps us with management and breeding decisions for our goats is to think about what animals we would like to see in our pastures in three, five, and ten years. As you read this book and the goat magazines, keep your eyes and mind open to what pictures make you think "wow." If the sight of your goats running toward you makes you happy, that's terrific. If the sight also makes you marvel at what amazing creatures they are, even better.

Our goat enterprise is profitable, which allows us to keep going, but more rewarding on a daily basis is the joy of renewal during each **kidding** season and pride in the goats that do well with their new kids, and the excitement when an animal is born that looks like what we have been trying to create by turning a specific buck in with a certain type or caliber of doe. Then customers call to find out when we will sell our market kids. Profit that lets us have pride in our animals and happy customers seems like a good cycle to me.

Goats will stand up on their back legs and harvest tree leaves as high as they can reach.

Excellent stewardship of the land and animals is evident on this West Texas Hill Country ranch. *Priscilla Ireys for Three Mill Ranch*

This could be someone's picture of the perfect place. A livestock guardian dog and her puppies enjoy a beautiful Canadian fall day guarding their caprine charges. *Pat Fuhr, Giant Stride Farm*

Conservation of a rare or endangered breed can be one of your goals in raising meat goats. This is a young female San Clemente Island goat, a critically endangered breed. *Carole Coates, Obsidian Ranch*

The Arapawa Island goat is a domestic goat that was once feral on an island near New Zealand. After hundreds of years of no interaction with humans, the critically endangered goats are now being conserved, and population numbers are growing slowly. *Jeannette Beranger, American Livestock Breeds Conservancy*

If it seems that this young South Carolina Island Spanish goat is smiling, it's because he and his herdmates have been moved to the mainland and no longer have to worry about alligators. *Jeannette Beranger, American Livestock Breeds Conservancy*

AN INDUSTRY OF OPPORTUNITY

The meat goat industry is unique, in that the demand for the product that we raise is purely based on consumer desire: There is year-round demand for fresh goat meat. The demand for *chevon* also doesn't require advertising, promotion, or free samples to maintain its growth.

Of course there are challenges in the market, but those offer incredible opportunities too. For example, a shortage of market goats in the winter causes prices to rise, while the ready supply of goats in the summer causes prices to fall. I see that as an opportunity to figure out how to have market goats available when the prices rise in the fall and winter. You can benefit from keeping your eyes open to new opportunities that align with your chosen goal(s).

Now that you've decided on a goal, whether it's helping the world or making some money, it's time to get ready for your meat goats.

Meat goats are very good listeners. They may not add much to the conversation, but it sure seems like they help you sort things out! *Beth Sladky for Hillcreek Farm*

THE PERFECT GOAT

The perfect goat is the one that does what *you* want or need it to do. This might be the biggest, meatiest goat that brings the best market price, or the landrace doe with the strong maternal instinct that produces three kids in one litter and raises them effortlessly and without assistance, or the 4-H meat goat project that won your daughter the blue ribbon at the state fair, or myriad other characteristics! The goats pictured on these pages all represent someone's idea of the perfect goat.

Red Boer buck. *Marge Newton, Newton Farms*

Loving doe and happy kid. *Horseshoe Canyon Ranch Meat Goats*

Child and goat both love to show. *4M Boer Goats*

Kalahari Red buck. *Albie Horn, Kalahari Reds Breeder, Hartebeeshoek*

Best buddies. *Childers Show Goats*

Goats eating kudzu, it's a beautiful thing. *Jean-Marie Luginbuhl, North Carolina State University*

Holiday finery. *Richard and Sandy's Boer Goat Farm*

Myotonic doe and kids. *Phil Sponenberg, Beechkeld Farm*

Buckskin-colored Spanish buck. *Chism Weinheimer, Weinheimer Ranch, Inc.*

A child gives a kid fresh greens. *Cori McKay for Smoke Ridge*

Dappled Boer doe. *Carol DeLobbe, Bon Joli Farm*

Pygmies are lovable. *Staci McStotts, Critics Choice Pygmy Goats*

CHAPTER 2

BEFORE THE GOATS ARRIVE

Getting your goat enterprise set up *before* your goats arrive is a great deal easier than trying to rush and put something together after the goats have gotten out of the trailer. This is especially true because some of the goats will be doing their best to "help" you with your project, and although they mean well, they aren't usually very good at building things.

What do you really *have* to have to begin a successful goat operation? Looking back with the clarity of hindsight, I imagined what I would have wanted in place before our goats arrived, and I created this handy checklist.

A word of caution when it comes to making purchases: Before you buy anything, first make sure you really need the item, figure out where the money

Having a secure pen to unload the goats into when they first arrive at your place is a necessity for their safety and your peace of mind. *Tangy and Matthew Bates, Blue Creek Livestock*

Meat goats are homebodies and, within a day or two of arriving at a new place, will stay close. If they have access to shelter, they will use it when they need to. *Priscilla Ireys, Critton Creek Farm*

Meat goats are very curious creatures, wanting to know what is going on in their world. *Michelle Merkel of Capricorn Acres*

will come from in your budget to purchase it, and then make a shopping list and stick to it. Don't give in to impulse buys! A store's job is to motivate you to spend as much money as possible. This is as true for farm stores as it is for toy stores or car dealerships. Spending your money is good for the store, but probably won't help you breed and raise better goats. Your job is to keep your eye on your goal and try to stay on the road that will get you there.

A secure pen or pasture—Whether the goats will arrive on a trailer, in dog crates, or on a semi, they are coming to a new and unfamiliar place. Your goats may not know what electric fencing is. Although an electric fence is a very effective and cost-effective way to keep goats where they should be, let your goats get settled in. After that there

Whether you will be using pasture, stored feed (hay), or some combination of the two for your goats, make sure that they get what they need, no more, no less.

will be time for them to explore and learn about **hot wire** if that is new for them. Chapter 4 covers many options for fencing your goats in (and keeping undesired visitors out).

Adequate and appropriate feed—Whether you will be using pasture, stored feed (hay), or some combination of the two for your goats, make sure that they get what they need, no more, no less. If feeding hay, giving the goats more than they need (for **maintenance**, growth, or for pregnancy and lactation) leads to wasting of hay, as the goats get picky and sort through their feed, discarding much of it. This wastes your money and makes it harder to break even or make you a profit in raising goats, in addition to making a mess that you will need to clean up before it fills up your barnyard. Keeping your goats underfed also causes problems, as hungry goats will put their considerable creativity and all of their efforts into escaping their enclosure. Underfed goats' ability to produce kids and milk to raise their kids will also suffer, which makes success in your goat endeavor that much harder to achieve.

Wisconsin offers excellent forage for raising meat goats. *Kim Hunter, Fossil Ridge Farm*

Minnesota's lush summer forage attracts the complete attention of these twin Boer kids. *LaDonna Hanssen Worsing, Prairie Road Farms*

These red Boers are happy and healthy in the warm and dry climate of Texas, especially with a round bale of hay at their disposal. *Texas Reds*

Clean, fresh water—The quality of the goats' water is extremely important, and yet very easy to overlook. When water stands still, algae soon blooms, and the goats become less inclined to drink readily. Ways to keep water fresh and clean are discussed in Chapter 5. If the goats will be able to drink from running water, such as a stream or river, check on the goats' access to the water and safety in reaching the water. If the goats will be drinking from a cattle-sized trough, it's important to make sure that they can reach the water easily (via steps or a graveled ramp of dirt) and also that they can get *out* of the trough if by some unhappy occurrence they get pushed in. Goats do many things well, but swimming isn't one of them.

In regions where forage is covered with snow in the winter, or when additional feed is needed, baled hay can give meat goats the roughage, protein, and energy that they need.

Clean, fresh water is a necessity for meat goats. Here, a pregnant Spanish goat drinks her fill for herself as well as for her unborn kids. *Freddie Brinson, Brinson Pineywoods Cattle and Spanish Goats*

Minerals—**Minerals** are good elements found in the earth for animals to consume, sort of like vitamins are for people. They're in the plants that goats eat as they forage in the pasture or in hay from a bale. But many areas of the country have smaller amounts of those minerals than what meat goats need to thrive. Most people raising meat goats find that it's very cost effective (usually only costing a few dollars per year) to have minerals available for their animals, either in granulated or block form. The goats eat what they need, and with the nifty PVC feeder shown in the photos on page 109 waste is minimal. Just keep in mind that if you're lucky enough to get rain frequently throughout the year, the feeder will need to be covered to prevent the mineral from getting soggy and solidifying.

Shelter—It seems that cold rain, or biting wind, is nasty weather according to goats. Even in frigid Montana winters, the goats will be outside, happily munching hay in subzero temperatures while the snow is falling. If it is above freezing but cold and rainy, the goats will find a roof to be under if they can. (The reason will be discussed in some detail in Chapter 4.) If the goats have trees or bushes in their pasture, or if they can get under an overhang or roof, they are much happier. Why does that matter? Because a happy goat is a healthy and productive goat.

Safety for the goats—Having your goats hurt or killed by coyotes, stray dogs, foxes, badgers, or

Most people raising meat goats find that it's very cost effective to have minerals available for their animals, either in granulated or block form.

WHAT'S A MENTOR? A BIG HELP!

There's just one more thing that will help you succeed with meat goats, and that is having a mentor. You can't buy one, unfortunately—you have to find or develop one. A mentor is someone who will let you bounce ideas off of him or her without laughing at you. Mentors may have been at it with goats for a year or a few decades, and they will help you learn why some things work and some things don't. And as you figure things out with meat goats, you might just make discoveries that your mentor can also benefit from.

What's great about mentors? Mentors are people who have some of the same wishes and hopes and goals in mind as you do for raising or producing or selling or loving or promoting meat goats. They might be folks that have (even just slightly) more experience than you. They might well be the people that you end up buying your goats from, and they might not be. Mentors care about your happiness and success. They may start out thinking that they just like you because you're a (potential) customer, and then they come to find out that you're a pretty neat person and it's good to know you. Am I speaking from experience? Absolutely.

mountain lions would be heartbreaking. It would also hurt your business, but it seems that to many goat owners the emotional cost of having animals for whom you feel responsible be maimed or killed far outweighs the financial loss. The numerous options for keeping your goats safe will be covered in Chapter 7.

If those six are the critical "have to have" items for you to start your meat goat operation, there will soon be other items that you *wish* you had. You could live without the wish-list items noted below, but living *with* them probably seems like it would be so much nicer, easier, more efficient, and more productive, right? You're correct; most of the tools on this wish list can make certain jobs easier and faster to do. Just be careful: It's much easier to take money out of your account than to put it back, and much easier to acquire things than it is to sell them again for a value anywhere near their purchase price.

On a beautiful spring day in Pennsylvania, these kids curl up in the corner of a shed while their mother is out eating. She will return soon to feed them. *Marcus Briggs, Dancing Heart Farm*

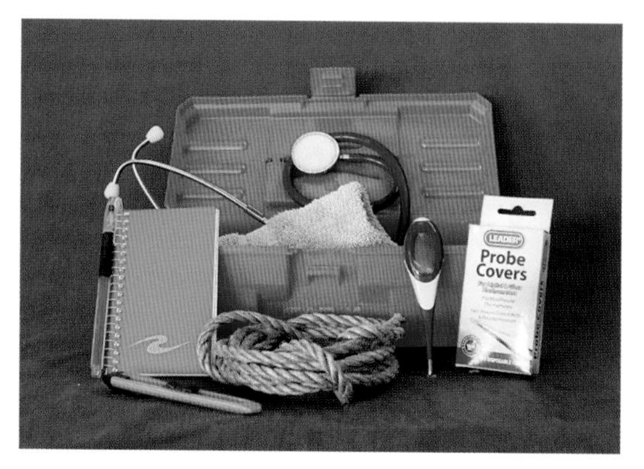

Kidding kit and tool kit.

Kidding kit—One fairly low-cost wish-list item that will move into the critical category as kidding season approaches is the kidding kit. The kidding kit contains everything you'll need during kidding season: a little notebook for recording births and any comments about the mother or kids, a pen, ear tags, a tagger, a bander, and bands (options for neutering are given in Chapter 8).

Tool kit—Your tool kit can be a small box or bag of any items you might occasionally need for your goats. It can include a lead rope, a thermometer (for taking a goat's temperature), a stethoscope, a hoof trimmer, and so on.

Working pen—A working pen or handling facility allows you to do something specific with or to your goats. A working pen can be as simple as one pen that you can crowd the goats into in order to **eye check,** deworm, vaccinate, or trim feet (if needed), or it can be a sophisticated facility with an alley, sort gates,

THE IMPORTANCE OF EAR TAGS

Why ear tags, you ask? Even if your herd will be made up of just a few does, if each doe has twins, you might not always be able to tell the kids apart by sight alone. Goats can change their appearance as they mature, and their coat colors fade or darken. Ear tags are probably the single most effective tool you can use in running your meat goat operation, and if you have more than a few does, accurate selection of keepers and culls is nearly impossible without individual identification. As your herd grows and your reputation for breeding excellent goats spreads, you might find yourself selling to buyers in other states. If that's the case, the goats will have to have ear tags to get the health certificate that allows them to travel to their new homes legally.

As your herd expands and you need to be able to judge the goats' weight accurately, a digital scale or a dial scale will become increasingly appealing.

and a raised working area that puts the goats at a comfortable level for the human handler. Even ranches that only handle their goats once or twice a year have some form of handling facilities. Sophisticated facilities can be custom welded, or simple working pens can be made out of combo panels wired to T-posts.

Scale—As your herd expands and you need to judge the goats' weight accurately, a digital or dial scale will become increasingly appealing. A scale never gets tired (unlike the human goat managers) and always tells the truth when it displays the goats' weight. How to use the information that the scale gives you to help pursue certain herd objectives will be discussed in Chapter 9.

Trailer—A trailer to transport your goats can be a wonderful help, but it can also be a fairly large

A three-sided shelter, open to the south, gives Spanish-Boer cross goats in West Virginia's mountains a place to hang out. *Priscilla Ireys, Critton Creek Farm*

This magnificent San Clemente Island buck appears to stand guard in the doorway of his barn in Nebraska. The guinea fowl offers parasite control by consuming ticks as well as serving as an alarm system. *John and Chad, Willow Valley Farms*

expense. If your enterprise will be concentrating on producing market kids, it's very likely that buyers will come to you to pick sale kids up. Some producers work together to organize trips where numerous goat operations share in the cost of the trip and all the goats get where they need to go. If you sell and deliver breeding stock, it becomes easier to justify buying your own trailer, but while it's easier to justify, it may not be much easier to pay for! And if you're going to pull a trailer, you'll need a truck that's able to do so. Having done the math more often than I would like to admit, I've determined that if you can find someone who's reliable and ethical and pay them or barter with them to make the deliveries rather than purchasing a trailer yourself, you'll often come out ahead.

SELECTING YOUR GOATS

Now that you have the six necessities (and maybe a wish-list item or two) ready for the goats, you get to do something really fun: deciding what kind of goats you want to get! A number of different kinds of meat goats are available for your consideration, as well as different ways to find goats to acquire. Some of these breeds you have probably heard of, or seen in pictures or in real life. Each breed or type of goat, as well as individual animal, has strengths and weaknesses. There is no perfect meat goat; each breed and every animal has *some* attribute that offers an opportunity for *some* improvement.

Actually, there is a perfect goat: the one that does what you want or need it to do for your

An Anatolian Shepherd guardian dog keeps watch over his herd of does and kids in Montana. His job is to make sure nothing disturbs his goats.

There is a perfect goat: the one that does what you want or need it to do for your goat enterprise!

goat enterprise! Take the paper with your goat operation's written goal, whether it's a phrase or a paragraph, and put it where you can easily see it as you look at the following pictures and descriptions. If you find a photo that resembles the picture of perfection in your mind, be it for your farm/ranch or the type of animal that you want your operation to produce, put it next to the paper. If you have a pack of small sticky notes, use them to mark the breeds that interest you.

Now please sit back, put your favorite beverage and snack within arm's reach, turn the telephone answering machine on, and let's take a tour around the United States. We'll be looking at different breeds of goats and what they have to offer the meat goat industry. The breeds are listed in the order that you might come across them, with no intent to convey ranking or grading. We'll start with the breeds that have been selected for their meat production, then look at goats that specialize in fiber production. There are breeds small in stature or populations and goats that are specialists in milk production. Since goats of different breeds will very happily and successfully breed with each other, composite or crossbred goats will also be discussed.

GOAT BREED SPECIALTIES

The Meat Specialists:

- Boer
- Kiko
- Spanish
- Savanna
- Myotonic
- Kalahari Red

The Fiber Specialists:

- Cashmere
- Angora

The Small Specialists (Physically or Numerically):

- Pygmy
- Nigerian Dwarf
- San Clemente Island goat
- Arapawa Island goat

The Milk Specialists: Dairy goats with large ears and with no ears.

The Blended Specialists: You can try to get the best of two breeds in one goat.

The white body and red head coloration is considered traditional by Boer producers. This doe appears proud of her kids, with good reason. *Childers Show Goats*

BELOW: The adaptability of meat goats to different environments and management styles is a great benefit to producers. Here, red and black Boers thrive in Texas. *Troy Powell, League Ranch*

The Meat Specialists

Breed: Boer

Description: A white, red-headed goat that produces meat, milk, and leather. Boer goats are tranquil, with a great meat-to-bone ratio.
Country of origin: South Africa
Average weight (mature doe): 170 pounds
Average weight (mature buck): 225 pounds
Average estimated productive lifespan (doe): 6 years
Average estimated productive lifespan (buck): 5 years
What people like best about the breed: easy keepers, meat production
What people wish they could change about the breed: quantity of milk production, udder structure, hardiness

An impressive red Boer in Washington enjoys a snack of alder pollen cones. *Richard and Sandy's Boer Goat Farm*

Boers can have many other colorations apart from the traditional, including solid red, solid black, paints, and white with black head, as demonstrated by this group in autumn pasture in Alberta. *Pat Fuhr, Giant Stride Farm*

This Kiko buck's coat is shiny, showing that he is in good health. To many producers, the color of the animal is secondary to performance. *Horseshoe Canyon Ranch Meat Goats*

Breed: Kiko

Description: The Kiko was developed for survivability and growth rate in the moist, temperate environment of New Zealand. Producers feel that Kiko does are excellent mothers and more resistant to internal parasites than many meat goats.

Country of origin: New Zealand

Average weight (mature doe): 130 pounds

Average weight (mature buck): 200 pounds

Average estimated productive lifespan (doe): 12 years

Average estimated productive lifespan (buck): 10 years

What people like best about the breed: parasite resistance, strong maternal attributes, very good foragers, foot health

What people wish they could change about the breed: Many meat buyers judge the amount of meat on the goat's frame by appearance, and sometimes downgrade the Kiko when it is offered as a market animal because Kikos can look rangier than a Boer or Boer-influenced goat.

This young Kiko doe is still growing, as evidenced by her hips being taller than her shoulders. *Goat Hill Kikos*

Kikos have a well-deserved reputation for parasite resistance. This young doe in Arkansas lives in a verdant environment and is thriving. *Horseshoe Canyon Ranch Meat Goats*

This Kiko buck respects the fence that holds him in his pasture and waits (more or less patiently) for breeding season. *Horseshoe Canyon Ranch Meat Goats*

The color pattern of this curious young Spanish kid is called "badger face." *Eddie Gonzalez, Yeates Farm*

Health and well-being are evident in this young Spanish buck in Wisconsin. *Lynne Shonyo, Shonyo Farm*

Breed: Spanish

Description: The Spanish goat is the product of natural selection. Goats that escaped or were turned loose by explorers and early settlers in North America from the mid-1500s on survived and adapted to their environment. Most are found in the arid Southwest, some in the humid Southeast, and many in the Midwest, Northwest, and Western states. The Spanish goats have shown their ability to survive and reproduce with absolutely no assistance or medications, but do respond very positively to an increase in feed quality or availability.

Country of origin: United States since the 1500s, Europe before that.

Average weight (mature doe): 95 pounds

Average weight (mature buck): 185 pounds

Average estimated productive lifespan (doe): 12 years

Average estimated productive lifespan (buck): 12 years

What people like best about the breed: hardiness, excellent mothering ability, longevity

What people wish they could change about the breed: the meat buyers' perception of the amount of muscling on the goat's frame and difficulty in taming older animals (Producers feel that Spanish kids raised around people are very easily managed when mature.)

This beautifully conformed young Spanish doe is looking to see what is approaching, and her kid is learning from her example. *Eddie Gonzalez, Yeates Farm*

Breed: Savanna

Description: a meaty, silver-white goat with dark pigmentation that is most visible in the horns, hooves, and nose

Country of origin: South Africa

Average weight (mature doe): 130 pounds

Average weight (mature buck): 210 pounds

Average estimated productive lifespan (doe): 8 years

Average estimated productive lifespan (buck): 9 years

What people like best about the breed: Savannas are excellent mothers, raising kids well with no assistance.

What people wish they could change about the breed: limited genetic base

RIGHT: On the other side of the world from his forbears, a Savanna buck thrives in West Virginia. *Priscilla Ireys, Critton Creek Farm*

Sheltered in the shade, these young Savanna does have no problem with the Texas heat. They are both patient and curious. *Priscilla Ireys, Critton Creek Farm*

The South African environment where the Savanna was developed is dry, and forage can be found growing on the branches of trees. Nonelectric wires in the fence are effective to hold these goats in a handling area. *Richard Browning Jr., Tennessee State University*

Breed: Myotonic

Description: Myotonic goats are unique in the goat world because when startled, their muscles lock up and the goat often falls over and lies still for a few seconds. They have typically been marketed as pets, due to their novelty, but are increasingly being sought after for use in breeding for meat production. The myotonic gene is exhibited only by purebred animals. Their meat-to-bone ratio is about 4:1, which is a dramatic increase over the 3:1 of nonmyotonic goats.

Country of origin: United States

Average weight (mature doe): 80 pounds

Average weight (mature buck): 135 pounds

Average estimated productive lifespan (doe): 10 years

Average estimated productive lifespan (buck): 10 years

What people like best about the breed: They have long life expectancy, are good mothers, have easy kidding and good milk production, are quiet (noticeably so), and are easy to manage because they cannot jump or climb. The breed's meat characteristics (tenderness and juiciness) are well liked by customers, and processors wish that more Myotonic or Myotonic-influenced goats arrived at their processing facilities.

What people wish they could change about the breed: the growth rate could be faster

This Myotonic doe and her daughter are lovely examples of the breed. Good conformation, attractive coat colors, and living proof of successful reproduction are all apparent. *Jeannette Beranger, American Livestock Breeds Conservancy*

The myotonic gene gives animals excellent musculature, as evident in this impressive buck in Wisconsin. Another beneficial effect of myotonia is that the goats do not jump. *Lynne Shonyo, Shonyo Farm*

Myotonic goats have typically been marketed as pets, due to their novelty, but are increasingly being sought after for use in breeding for meat production.

Since these goats are calm and content, there is no evidence of their myotonia, the trait that causes their muscles to lock up. This doe and her adorable kid are foraging happily on early spring growth. *Jeannette Beranger, American Livestock Breeds Conservancy*

Breed: Kalahari Red

Description: Solid red, the Kalahari Red has good foraging abilities, high fertility, and excellent reproduction rates. Many does consistently raise three crops of kids in two calendar years in current South African production.

Country of origin: South Africa

Average weight (mature female): 120 pounds

Average weight (mature male): 200–210 pounds

Average estimated productive lifespan (breeding female): 6–8 years

Average estimate productive lifespan (breeding male): 6–8 years

What people like best about the breed: strong maternal instincts, color, fertility, and **reproductive efficiency**

What people wish they could change about the breed: Meat goat producers who raise and love Kalahari Reds say, "Why change something perfect?" (Author's note: As of the writing of this book, Kalahari Reds have not yet been imported into the United States or Canada.)

A group of Kalahari Red does show how their color makes them less visible to predators. *Albie Horn, Kalahari Reds Breeder, Hartebeeshoek*

Although meaty, this Kalahari Red doe shows femininity in her head and neck. The angulation in her **hock** (the joint in her back leg) allows her to travel efficiently over distances. *Albie Horn, Kalahari Reds Breeder, Hartebeeshoek*

While well proportioned and smoothly conformed, this Kalahari Red buck shows impressive musculature in both his forelegs and rear legs. *Albie Horn, Kalahari Reds Breeder, Hartebeeshoek*

The magnificent curving horns on this buck in Ohio have been growing since birth. You would be equally correct in calling him a Spanish or a Cashmere buck. *Dian Morefield, Morefield Spanish Goats*

The Fiber Specialists

Breed: Cashmere

Description: Cashmere has recently been recognized as a breed—the North American Cashmere goat. Cashmere is the name of the fine underdown grown by any goat other than an Angora. Goats grow cashmere from the summer solstice (late June) until the winter solstice (late December) and then begin to shed the fiber. The fiber is harvested by combing or shearing. To qualify as cashmere, the fiber must be less than 18.5 microns in diameter (roughly one-third the thickness of a human hair) and have "crimps" rather than the fibers being straight. Crimps are tiny waves or crinkles. Goats can be selected for increased cashmere production.

Country of origin: Goats growing cashmere were first identified in the Himalayan mountains of India, where they were originally called Kashmir goats.

Average weight (mature doe): 115 pounds

Average weight (mature buck): 175 pounds

Average estimated productive lifespan (doe): for cashmere and offspring production, 12 years

Average estimated productive lifespan (buck): for cashmere production, 4 years; for kid production, 10 years

What people like best about the breed: the super-soft, luxurious fiber and the antics of the lively kids

What people wish they could change about the breed: If the goats have the ability to do so, they can get very creative about deciding where the boundaries of their pasture are.

Black Cashmere goats, like this beautiful buck, produce brown cashmere. The luxurious fiber is prized by handspinners as well as those in the garment trade. *Black Locust Farm*

The Cashmere goat is now recognized as a distinct breed, although any goat other than an Angora can produce cashmere, the fine fibers grown by a goat's secondary follicles, from summer solstice until winter solstice. This young buck in Maine enjoys summer's bounty. *Wendy Pieh, Springtide Farm*

Curiosity about the photographer temporarily causes a white buck to stop browsing on lush Maine forage. *Wendy Pieh, Springtide Farm*

Cashmere is the name of the fine underdown grown by any goat other than an Angora.

At a breeding stock sale, an evaluation of this buck's fiber is shown on the small chalkboard above his back. *Joe David and David Ross, Ross Ranch*

Breed: Angora

Description: The Angora goat produces mohair, a fine fiber with a soft "hand" and natural luster. Mohair is prized in the production of both clothing and upholstery fabrics, and it is increasingly sought after by hand-spinners.

Country of origin: Turkey, after whose capital city (Ankara) the goats were originally named

Average weight (mature doe): 95 pounds

Average weight (mature buck): 175 pounds

Average estimated productive lifespan (doe): 9 years

Average estimated productive lifespan (buck): 9 years

What people like best about the breed: the luxurious fiber

What people wish they could change about the breed: Many breeders are moving toward a broader set of animal production parameters than exclusively selecting for high quantities of fine fiber production.

Angora goats have been selectively bred for centuries for their luxurious fiber. With only one type of follicle, the Angora produces mohair, which must be shorn annually. *Karen Stieler*

This Angora kid seems to be saying, "There's a big world out there, and it all needs to be explored." *Karen Stieler*

Most baby animals are attractive, and these Angora kids are no exception. They have been ear-notched for identification purposes. *Joe David and David Ross, Ross Ranch*

The Small of Stature or Small in Numbers Specialists

Breed: Pygmy

Description: A small goat with a good-natured personality, Pygmies are known for their friendliness and hardy constitution.

Country of origin: Cameroon, West Africa

Average weight (mature doe): 50–75 pounds

Average weight (mature buck): 60–85 pounds

Average estimated productive lifespan (doe): 10–15 years

Average estimated productive lifespan (buck): 10–15 years

What people like best about the breed: their small size and pleasant personalities

What people wish they could change about the breed: their creativity in escaping confinement

Pygmies are gregarious goats, stocky and compact. They are often kept as companion animals. *Staci McStotts, Critics Choice Pygmy Goats*

No sibling rivalry here, as these twin Pygmies enjoy each other's company as well as the Wisconsin sunshine. *Kim Hunter, Fossil Ridge Farm*

Breed: Nigerian Dwarf

Description: The Nigerian Dwarf is a miniature goat of West African origin. Its build is similar to that of the larger dairy goat breeds, and the parts of its body are in proportion. The nose is straight and ears are upright. The coat is soft with short to medium hair. Any color or combination of colors is acceptable.

Country of origin: Those pictured are American Nigerian Dwarf goats.

Average weight (mature doe): 50–60 pounds

Average weight (mature buck): 60–75 pounds

Average estimated productive lifespan (doe): 11 years

Average estimated productive lifespan (buck): 11 years

What people like best about the breed: They like their size, personality, gentleness, and milk production. The Nigerian Dwarf is a working pet and a great all-around goat.

What people wish they could change about the breed: The only trait that one owner of Nigerian Dwarfs wished she could change was that she thought her goats occasionally felt themselves superior to their owner.

The Nigerian Dwarf doe is well known for her excellent milk production. The breed's moderate size and gentleness make it a wonderful "working pet." *Four Zs Miniz, Jack and Nicki Zoda*

San Clemente Island goats are loved by their owners for their kind and gentle dispositions as well as their deerlike beauty. *Leslie Edmundson, Little River*

This young San Clemente Island kid will keep an eye on the photographer while Mom naps. *Megan and Damon Burt, Fraggle Rock Farm*

Breed: San Clemente Island Goat

Description: A critically endangered rare breed, the San Clemente Island goat was feral on an island off the coast of California for more than a century.

Country of origin: the United States

Average weight (mature doe): 70 pounds

Average weight (mature buck): 100 pounds

Average estimated productive lifespan (doe): 12 years

Average estimated productive lifespan (buck): 12 years

What people like best about the breed: Small and deerlike, with kind and gentle dispositions, their beauty is breathtaking.

What people wish they could change about the breed: They wish there were more of the goats; owners are helping the goats as best they can to remedy the shortage.

Breed: Arapawa Island Goat

Description: A feral breed of domestic goat, the Arapawa (pronounced *ar-a-PA-wa*) sprung from ancestors that arrived with European colonists in New Zealand as early as the 1600s. The goats were rescued from a planned government eradication program in the 1970s. Although still currently listed as critical by the **ALBC** (the American Livestock Breeds Conservancy), Arapawa Island goat numbers are slowly increasing in New Zealand, the United Kingdom, and the United States.

Country of origin: New Zealand

Average weight (mature doe): 70 pounds

Average weight (mature buck): up to 125 pounds

Average estimated productive lifespan (doe): 6–8 years

Average estimated productive lifespan (buck): 6–8 years

What people like best about the breed: They are a medium-sized, all-purpose milk and meat family goat. They bear their young without assistance and are presumed to be relatively disease-free.

What people wish they could change about the breed: They'd like a broader genetic base for the U.S. herd, as there were only six animals in the founding herd.

The Arapawa Island goat lived on an island off the coast of New Zealand as the feral offspring of goats that had come with English settlers hundreds of years earlier. *Jeannette Beranger, American Livestock Breeds Conservancy*

The Milk Specialists

Breed: Nubian

Description: The Nubian (also known as the Anglo Nubian) is a dairy goat known for its high-butterfat milk, which is prized by cheesemakers. Because of their fairly large size, Nubian goats can also be considered an all-purpose goat, useful not just for their milk, but also for meat production, especially when crossed with meat goat breeds such as the Boer. The Nubian has an aristocratic bearing, with high head carriage, a Roman nose, and very long, pendulous ears that hang close to the head.

Country of origin: Modern-day Anglo Nubians were developed in England during the 1800s by crossing English goats with goats of Indian and African origin.

Average weight (mature doe): 150 pounds

Average weight (mature buck): 180 pounds

Average estimated productive lifespan (doe): 10–12 years

Average estimated productive lifespan (buck): 8–10 years

What people like best about the breed (for use in a meat goat–breeding operation): Nubians can be bred year-round. Nubians also produce enough milk to foster additional kids.

What people wish they could change about the breed (for use in a meat goat–breeding operation): The easy-going nature of the Nubian can be a disadvantage when Nubians are in a herd with more assertive meat goats.

The Nubian (also known as the Anglo Nubian) is a dairy goat known for its high-butterfat milk, which is prized by cheesemakers.

The Nubian dairy doe can give extra milk if needed by a meat goat producer in the rare event that there is an orphaned kid. *Tangy and Matthew Bates, Blue Creek Livestock*

Other Dairy Goats

Other breeds of dairy goats, including the Saanen, Toggenburg, Alpine, Oberhasli, and La Mancha, can offer meat goat producers who crossbreed their goats with "the Milk Specialists" the addition of genetics for higher milk production. Remember that dairy goats have been selected for generations to convert foodstuffs to milk, not meat, but dairy goats play an important role in some meat goat enterprises.

Even dairy goats will choose taller forage over short grasses, as this Saanen doe in Ohio shows. This is how goats have avoided contamination by parasites for millennia. *JS Family Farm*

Nubian-Myotonic
If a goat is the result of breeding a Nubian to a Myotonic, should it be called a Myonian or a Nubatonic? Most processors would simply call this kid very attractive. *Lynne Shonyo, Shonyo Farm*

Breed: Composites

Any breed of goat can, and if given the opportunity will, breed with a goat of the opposite sex. Meat goats are not usually snobs and do not appear to care what lineage and pedigree the goat of the opposite sex has. In researching available meat goats, you may read of

- Boki (Boer–Kiko composite)
- Sako (Savanna–Kiko composite)
- Sabo (Savanna–Boer composite)
- Genemaster (3/8 Kiko–5/8 Boer)
- TexMaster (Tennessee Meat Goat and Boer)
- Tennessee Meat Goats

Since you will have your goals and objectives defined for your meat goat enterprise before you start goat shopping, one or more of the composite breeds may well offer you what you're looking for.

What is the difference between a composite and a cross-bred? Another meat goat producer told me recently (not completely jokingly) that a composite is the result of planning, while a cross-bred is not. If your objective is to raise healthy, profitable meat goats, an argument could be made that

If your objective is to raise healthy, profitable meat goats, an argument could be made that purity of bloodline isn't as critical as performance.

A GOAT BY ANY OTHER NAME: BREED CONSERVATION

You may well come across the terms *breed conservation* and *genetic diversity* as you learn more about meat goats. Wait! Please don't turn the page. We as meat goat breeders have a living, breathing example of why breed conservation matters, why we shouldn't let all of something slip away just because we can't see the animal's value right now.

The Spanish goat survived (some would say prospered), usually without any help from people, since coming to the Americas in the 1540s. They were just part of the landscape in Texas and the southwestern United States. When the Boer goat came to North America in the early 1990s, many Spanish does were bred to Boer bucks. After their beautiful crossbred kids were born, many of the Spanish does were sent to market because they didn't look as impressive as their offspring. After a few decades, there are now Boer and Boer-cross goats from coast to coast and only a fraction of the previous number of Spanish goats. What if those tough little goats had something to offer to meat goat producers? Luckily, some Spanish goat breeders refused to change to the new and improved meat goat, and we can still go back to the well and get some Spanish blood if it will help us build a better meat goat for our environment, customers, and management system.

Remember also that when it comes to breeds, "better" is a matter of opinion. Only you can decide what *you* want and need for your enterprise to survive and thrive.

purity of bloodline isn't as critical as performance. If your objective is to raise pedigreed, registered breeding stock, then the purity of bloodline is very important. Remember: the perfect goat is the one that does what you want or need it to.

Since the **generation interval** in meat goats can be as short as twelve months (or less), trial breedings of different crosses can be carried out quickly and the results of the matings seen and compared to the results of other matings. If there was such a thing as a goat census, the results from a hundred years ago, the results from today, and the results from a hundred years into the future would probably look quite different. Agriculture has certainly modified cattle, pigs, and sheep in recent decades, and there is no reason to think that meat goats will not also be selectively bred to more aptly fit mankind's needs.

Spanish-Nubian
When the innate hardiness of the Spanish goat and the size and docility of the Nubian are blended, the composite goats resulting from the breeding are beautifully conformed and also very attractive! *Kim Hunter, Fossil Ridge Farm*

CHAPTER 3

BUYING AND TRANSPORTING GOATS

Now you've set your goals, gotten your place ready, and finished reading the breeds section of Chapter 2. Good for you! If there were a report card for all of this, you'd have an A+.

Now you get to go buy some goats.

KNOW YOUR ABCS: AUCTIONS, BREEDERS, AND CHOICES

Of course I'll tell you that there is an *easy* way to buy meat goats, and a *good* way to buy meat goats. Once again, you have to start smart. Don't give in to the temptation of competing with anyone and buying more animals than you can comfortably handle. That might be five, fifty, or five hundred goats, depending on your plans and the time and money you have available.

Auctions

Many goat producers and goat magazines will advise you *not* to buy goats from a sale barn. They're not just saying that because they might have goats for sale, and they hope that you'll buy your goats from them instead of their competition. They also say it because your chances of getting productive and profitable goats at a sale barn (also called a livestock auction) are usually not as good as when you buy the goats directly from another producer. The livestock auction is where people take their animals to sell them, typically as market animals, for processing and consumption (slaughter). There might be a terrific goat for sale at "the ring," with good fertility, a healthy, functional udder, and solid maternal attributes so she'll want to raise her kids. But she may have been taken to the sale because she isn't fertile, or her kids don't grow well (or at all). On the plus side, sale prices at a livestock auction may typically be lower than those of a successful producer. Unfortunately, if you have questions about a goat that you bought at the auction, you can't just call the seller to get more information.

Some auction facilities hold special breeding stock sales, which can be an opportunity to purchase

Learning how the owner calls the animals to come in, and seeing how the goats interact with each other, will help you manage the animals once they are on your property. *JS Family Farm*

By visiting a producer's farm before buying any goats, you can see how the animals are cared for and how they interact with the people who raised them. *Kim Hunter, Fossil Ridge Farm*

quality goats. If you're thinking about attending a breeding stock sale, find out as much as you can about the animals that will be sold beforehand. Do they come with production records or information about the person who consigned them to the sale? As with any purchase, the better informed you are as a buyer, the more likely you are to end up with an animal that works for you.

Breeders

Another option is to go to a breeder for your goats. When you buy a goat from a breeder, you're buying the years that the breeder has invested in developing the animal. Unlike an auction, the person selling goats directly has much more to offer than just the animals, although he or she may not know it.

You may find a breeder by spotting an ad in a goat magazine. You might come across a breeder's website as you're browsing online. Either way, keep in mind that not everyone selling good goats has a fancy sales packet or a beautiful website. All that's required is that you can find them, that you can learn about what animals they have for sale, and that you can receive courteous and detailed answers to your questions.

Choices

Now that you've decided where to buy your goats, how do you choose between the ones available?

If you're buying from breeders, ask them what they like best about their herd. Find out what traits they're working on improving, and why.

Some people carry the index card with the picture of their perfect goat; they may have written their goal phrase on the reverse side. Since I'm such a believer in cheat sheets, my index card also includes the top three things that the perfect goat in the picture is going to do for my herd.

You can evaluate firsthand if the goats could be described as "head up, tail up, bright-eyed, and interested," which indicates health and happiness. *Renard Turner, Vanguard Ranch*

If you're buying from breeders, ask them what they like best about their herd. Find out what traits they're working on improving, and why. Ask to see the records for the goat(s) you're thinking of buying and have the sellers explain why they think those goats would be good for you. Remember, if that seller doesn't want to offer you this information, many other sellers will be glad to. Of course, this doesn't mean you should expect unlimited free advice: Always respect the seller's time. Know what you're looking for (or at least, what you *think* you're looking for) and listen carefully to what the seller says in response to your questions.

Which goats you buy *will* affect your future, so it's important to be honest with the seller and yourself about your wants and needs. For example, if you want to enter your goats in beauty pageants (no, I don't mean shows), then tell the sellers so; they will likely direct you toward the prettiest goats they're selling. Similarly, if you want to enter your goats in size competitions, let the sellers know so they can show you their biggest goats.

If you simply want to raise healthy and productive goats, ask the sellers which goat they wouldn't part with at any price, then ask if there are any relatives of that goat that *are* for sale. If you want to raise profitable goats, ask the sellers which of their goats raise the most pounds of kids per pound of mother goat and pound of input (acres of pasture and pounds of hay or grain). The seller probably wouldn't sell you that goat, but again, maybe that goat has a son or daughter that might be for sale.

When shopping for your goats, don't forget that they'll need to get from where they are to your farm or ranch, which may be down the road or a long way away. Some sellers can deliver the goats themselves or make arrangements for delivery. Be sure to ask about your options, and don't forget the details: How much would it cost to have the goats delivered? How would the goats be transported? When could you expect them to arrive? Depending

If the goats are handled calmly and quietly at the producer's farm or ranch, you will be able to do the same at your place. *Morgan Frederick for Three Mill Ranch*

on the answers to your questions, you might decide to have the goats delivered, shop around for other delivery methods, or pick up and transport the goats yourself. (I'll explain more about traveling with goats later on in this chapter.)

One final warning before you hand over your money: Size can be misleading when it comes to goats. Goats that are born and raised as a single birth will usually be larger per day of age than a kid that was born and raised from a twin, triplet, or quad birth. Where a single kid gets all of the nutrition in utero (growing inside its mother), and all of the milk once it is born, kids from multiple births have to share the dinner table with their siblings, and thus usually don't grow quite as quickly as their single counterparts. However, if you're just interested in your goat enterprise's bottom line, sheer size of any given kid might not matter much: A doe that has two slightly smaller kids and raises them without help will end up giving you more goats that you can sell to buyers as individual animals ("by the head"), and it will almost always give you more total pounds of kids to sell, if that is how they are sold.

Savanna-Spanish and Boer-Spanish cross goats nibble on fall leaves and twigs in West Virginia. *Priscilla Ireys, Critton Creek Farm*

FOR WHAT IT'S WORTH: COST AND VALUE

The value of a goat is whatever the buyer and seller agree on, meaning that a goat's true value (and consequently its price) can often be arbitrary. That's why having a written goal, timeline, and budget is so important. With those tools, you can better judge whether the goat you just fell in love with is worth the price and if it can and should be a part of your future.

As when shopping for most things, a high price on a goat isn't always a guarantee of quality, and a low price doesn't mean that a goat isn't worth buying. Whether high or low, you should always ask why the seller has set the price on the goat in question, and why he or she feels that it's fair. Listen to the answer with an open mind and decide for yourself what the goat is worth to you. To help you with your decision, I've included two worksheets for calculating a goat's true value (one for does, one for bucks).

The value of a goat is whatever the buyer and seller agree on, meaning that a goat's true value (and consequently its price) can often be arbitrary.

This lovely Boer doe in Idaho seems to be looking calmly but with curiosity at what is over there. *Simon Boers and Capra Chevon*

THE VALUE OF A PERFECT GOAT: DOES

Purchase price of the goat:		________
Cost to get the goat home:	+	________
Cost of goat at your place:	=	________

Cost of goat at your place:		________
Expected productive years of life left:	÷	________
Cost of goat per year:	=	________

Cost of goat per year:		________
Annual maintenance cost (feed, water, minerals, share of the herd vet bills, ear tags for goat kids, share of guardian animal cost):	+	________
Total cost of animal and maintenance per year:	=	________

Income Estimate for Production of Market Animals, Sold by the Pound

Estimated number of kids raised for sale (in each of the estimated years left of production):		________
Estimated weight at sale:	x	________
Estimated sale price per pound:	x	________
Costs of transportation to the sale and commissions:	–	________
Estimated gross income per year:	=	________

For Production of Breeding Animals, Sold by the Head

Estimated number of kids raised for sale (in each of the estimated years left of production):		________
Estimated value at sale:	x	________
Costs of transportation to the buyer and commissions:	–	________
Estimated gross income per year:	=	________

Either estimated gross annual income figure:		________
Total cost of the goat per year:	–	________
Net (pretax) income per year:	=	________
Estimated lifetime left (in years):	x	________
What that goat might net you before you have to decide how she'll leave your breeding herd:	=	________

THE VALUE OF A PERFECT GOAT: BUCKS

This example is only to show you how *we* price animals, and it is not intended as a recommendation. You may well be able to find a good buck that will help you achieve your goals for less than we charge, and other producers may charge more.

Market value of the number of kids that the buck will sire in one year:		________
.02 (2%) (Please note, this is an arbitrary number to use for valuation):	x	________
Price for which we sell a buck:	=	________

Price of buck:		________
Estimated lifetime left (in years):	÷	________
Total cost of animal per year and maintenance per year:	=	________

(Remember to add what the buck will cost you annually before you decide how he will leave your breeding herd).

HAVE GOAT, WILL TRAVEL

Now that you've chosen your goats, there's just one small detail to take care of: getting them to their new home!

If you've chosen to have your goats delivered, you can rest relatively easily. If on the other hand you've decided to transport the goats yourself, you'll have a number of steps to take to ensure that the goats get to their destination in good condition. To help you in this process, I've answered the most common questions people have about transporting goats.

Are there special vehicles for transporting goats?

Lots of vehicles can be used to transport one or more goats: dog crates, goat totes carried in trucks or vans, pickup toppers, trailers, and semis have all been used to move goats from one place to another.

An alley leading away from a holding pen will make it easy to load the meat goats into a trailer, pickup topper, or goat tote. *Tangy and Matthew Bates, Blue Creek Livestock*

How long can goats travel?

So long as goats are protected from extremes of temperature (heat *and* cold), have access to fresh air, have enough room to stand up and lie down, and are fed and watered each day, they can easily take trips up to several days in duration. Making sure that the motion of the vehicle is as gentle as possible (smooth speeding up and slowing down) will make the trip much easier overall for the goats.

Lots of vehicles can be used to transport one or more goats: dog crates, goat totes carried in trucks or vans, pickup toppers, trailers, and semis have all been used to move goats from one place to another.

How many goats can travel in a dog crate, topper, trailer, or semi?

Unfortunately, there's no very simple answer to that question. It depends on the size of the goats and the size and strength of the conveyance.

Here is a *very* rough formula for calculating how many square feet of floor space you'll need per pound of goat to be transported: Take the number of animals and multiply that by the average weight of each animal. The resulting number is the total pounds of goats to be carried. Divide that number by thirty-five, and you will have the *approximate* number of square feet of floor space needed. Divide that number by the width of the vehicle and you will have the necessary vehicle length.

For example, let's say you want to transport five 150-pound does. A friend has a 12-foot-long trailer that's 6.5 feet wide inside and is willing to go with you to get your goats. Your friend asks you if the goats will fit in the trailer. You grab this handy cheat sheet and a calculator (or pen and paper) and figure out that

5 x 150 = 750

750 / 35 = 21.4 square feet

21.4 / 6.5 = 3.3 feet (This means you'll need only 3.3 feet of the 12-foot trailer, so yes, the five goats will all fit easily.)

When transporting bucks, remember to take their horn span into account. You'll also want to plan for their desire (or lack thereof) to be in close proximity with other goats, as well as the desire (or lack thereof) of the other goats to be in close proximity to them.

In addition, don't be fooled if someone says, "Oh, you need only so much space for a doe." Your advice giver might not be thinking of the same breed that you're transporting—a mature Nubian (usually considered a dairy breed), for example, might easily need twice as much space as a Spanish doe, because the Nubian is twice as big and heavy.

How will the goats get onto the trailer?

If the sellers have sold goats before, they should be set up to get your goats safely loaded into your trailer or topper. When you arrive, the goats you're buying should be in a separate pen with a gate or alley to your trailer, unless you or the seller

A "goat tote" is a portable crate that fits in the back of most pickup trucks. In regions that have cold temperatures or precipitation, a secure cover would be needed to safely transport meat goats. *Carol DeLobbe, Bon Joli Farm*

Plastic crates designed to carry dogs on airplanes can be used to comfortably transport goats. The crates come in different sizes.

However it's done, loading your goats should be a calm and pleasant experience for the goats and for you.

can walk each animal into it individually. Loading alleys and ramps are a wonderful help, but straw bales set up like stairsteps work just as well for allowing goats to climb aboard. Panels can also be set up as a temporary fence to give goats a path to the trailer. Just make sure they're strongly secured—don't trust baling twine or stretchy cords to hold your panels in place.

However it's done, loading your goats should be calm and pleasant for the goats and for you. Both of you have a trip ahead of you (possibly a long one), so it's good to start out on a high note.

Can I double-deck a trailer and carry more goats?

Yes, you can. Some trailer manufacturers make double-decker trailers, which are set up to safely transport two layers of animals. You can usually remove the upper deck if you're transporting larger animals like cows or horses. If you install a do-it-yourself upper deck in a regular stock trailer, remember that the goats riding on the lower deck need to be able to breathe. With two layers of goats on board, you might exceed the maximum **carrying capacity** of the trailer's axles. The pulling power *and stopping power* of the vehicle that you have a trailer hooked to should be more than what you'll need.

If I have a group of goats loose in a trailer, should I divide them up?

Goats are very good at squeezing into small spaces, which can be detrimental to the smallest goats in a group. For example, sixty-five kids weighing fifty-five pounds each will fit very comfortably in an eighteen-foot-long trailer. But it would be safest for the kids if you divide them into two groups and use the center divider of the trailer to keep the groups separated. A group of full-grown does and some kids that are no longer nursing would travel more comfortably and safely with each age group in their own compartment.

What's the best kind of bedding to use in a trailer with goats?

Bedding is used for absorbing urine and so that the goats don't have to lie down directly onto a wet or cold floor. Some people prefer to use wood shavings as bedding, while others like to use straw. Other handy things to bed your trailer/topper/dog crate may be available in your area. The major factors in deciding which bedding to use are availability, affordability, functionality (how well it does its job), how easy it is to clean out, and what you can do with it after cleaning it out. For our operation, we've found that after a trailer trip with goats on board, wood shavings work quite nicely as garden mulch.

How often should I feed and water the goats while traveling?

Many goats will contentedly eat hay while their conveyance is moving and drink when stopped (for example, if you stop for fuel or a meal). On a longer trip, should you stop for the night, or try to keep rolling and get to your destination? If you have a copilot, and you can trade off driving and sleep while the other person drives, getting home safely in the shortest possible time will usually be easiest on the goats. Putting **electrolytes** in their water (which you should offer them every eight to twelve hours) will give them an extra boost during and after the trip. For goats, drinking electrolytes is like you drinking a health tonic. Electrolytes also have an added benefit: They can make unfamiliar water taste better to your goats.

If we arrive after dark, can I unload the goats before morning?

Remember the six things you needed before you brought your goats home? The very first item on the list is a secure pen to unload your new goats into. Not a big pasture with electric fencing, even if the goats are used to hot wire, but a smaller area with a secure physical fence, or a pen made out of rigid wire panels. A water source should be in the pen and a shelter so the goats can get out of any wind or rain.

Even if the goats are used to guardian dogs, if you have a guardian dog or a house pet, it will be new to them. Keep the dog outside the pen. If the goats are not yet used to guardian dogs, definitely keep the dog well away from the pen until it is daylight. When everybody has had a chance to settle in and get acquainted through the fence (which could take a few hours or a few days), you should be present to manage the interactions when you do face-to-face introductions.

A stock trailer can be double-decked to increase the floor space that can carry meat goats. The goats on the lower level must have ventilation. A portable ramp makes loading and unloading simple and safe. *Weed Goats 2000*

My new goats keep shoving and butting each other. What's wrong?

Now that your goats are home, you're probably pretty excited to start learning about them: what they're like, who's the head goat, and which goat wants to be your friend. But before you do, give your goats time to figure out who's who. Even if your new goats all come from the same place—and especially if they came from different places—they're going to have to reorganize the herd structure. And this will happen no matter the size of the herd; even just two goats will establish a pecking order. So expect some gentle (or not-so-gentle) pushing and shoving at first as the goats establish who's the boss and who gets bossed around.

CHAPTER 4

FENCING AND FACILITIES

Having your goats stay where they should be, go where you want them to go, and thrive no matter what the weather offers will make your experience with meat goats a pleasure, making all of your goals easier to achieve.

Items for fencing, handling, and sheltering your goats can most certainly be gotten without breaking the bank, or your back muscles. Let's look at how some successful meat goat producers have done just that.

Putting goats of opposite sexes on either side of a fence is a good test of how well-constructed the fence is. This fence in Idaho is tall enough that the goats have never thought it possible to go over it and so don't try. *Tangy and Matthew Bates, Blue Creek Livestock*

DO FENCE ME IN

Goats have a reputation for being notoriously difficult to fence in. Some goats can be, but there's usually a logical reason for the behavior. For example, if a goat is hungry or bored, or has lived its life without any effective confinement, it will use its impressive creativity to get to the food, fun, or freedom on the other side of the fence.

So what can you do to keep your goats secured? There are several popular methods: physical fencing (such as panels or woven wire fences), electric fencing, distance, and staggered grazing schedules. This chapter will explore each of these options in greater detail, but before that, one piece of advice: Check the property and zoning rules for your locale before you purchase fencing. In some states, the fence around a property's outside perimeter by law must be a physical fence. For your peace of mind, find out what rules you need to follow to keep your goats secure without running afoul of local ordinances.

Physical Fencing

When it comes to physical fences, less can often be more. If goats have adequate food, shelter, and safety from predators, your fence doesn't have to be expensive, prison-grade material to be effective.

One popular variety of physical fencing is a flexible metal fence called *field fence*, *woven wire*, and *page wire*, depending on the area. Physical fencing is more expensive than its electrical equivalent but rarely requires upkeep.

Rectangular openings in the fence allow horned meat goats to remove their heads easily after reaching through for a tasty morsel on the other side. *Dr. Ken Andries, Kentucky State University Land Grant Program*

One of the few criticisms of physical fencing is that when meat goats push their heads through the fence to reach for something on the other side, their horns can make it difficult or impossible for the goats to free themselves. Some producers solve this problem by dehorning or **disbudding** their goats. Others use a simple and inexpensive device commonly known as a **horn blocker**. Still others use an electric wire set approximately eight to ten inches off the fence to prevent goats from getting

If goats have adequate food, shelter, and safety from predators, your fence doesn't have to be expensive, prison-grade material to be effective.

close enough to put their heads through the fence in the first place.

One of the easiest ways that many producers have found to keep goats from getting stuck is to use a woven wire fence with rectangular openings. Since the openings in the fence are wider than they are high, and goats typically turn their heads sideways to try to get through a fence, stuck goats can easily free themselves without any help.

If you're thinking about keeping your goats contained with a four- or five-strand barbed wire fence, think again.

If you're thinking about keeping your goats contained with a four- or five-strand barbed wire fence, think again. A cow might balk at such an obstacle, but to a goat, there's plenty of room to escape below the bottom strand of wire. A single electric wire **offset** about eight to ten inches away from the fence on the goat's side, and the same distance up off the ground, effectively blocks the goat's path. How can just one wire do that? Most meat goats will try to crawl under a fence, so the

This woven-wire fence will never be rubbed on by a goat or have a goat head pushed through it due to the electric wire offset on the inside of the goat pasture. *Marcus Briggs, Dancing Heart Farm*

This goat facility is subdivided into various pastures, which allows different groups to be kept separate or the entire herd of meat goats to be rotated between pastures to minimize parasite contamination. *Liberty Pastures*

area that they see as the way out is the bottom ten inches. If there is an offset electrical wire, when the goat gets down to crawl under the fence, it will come in close contact with the earth, which grounds the wire's electrical charge. The resulting shock, while harmless, will be very memorable to the goat.

A few goats simply refuse to be contained and are liable to jump over a fence. Speaking from personal experience, and with an admitted bias, I can only say this: If a goat's a jumper, it had better have some very impressive traits to outweigh that behavior, at least if it's going to live with us!

Electric Fencing

Electric wires use electrical shocks to keep goats contained because the shocks convince the goat that the fence should be respected. Hot wire fences can be energized by a fence charger that's plugged into an electrical socket, or by a solar charger that gets power from the sun's rays. Please see page 77 for a simple diagram of how an electric fence functions.

To test your fence, use the nifty little tester that you acquired with your fence charger. Lay the ground spike on the earth, step on it, and lay the knob of the tester on the energized wire.

"Composite" fence

Ineffective

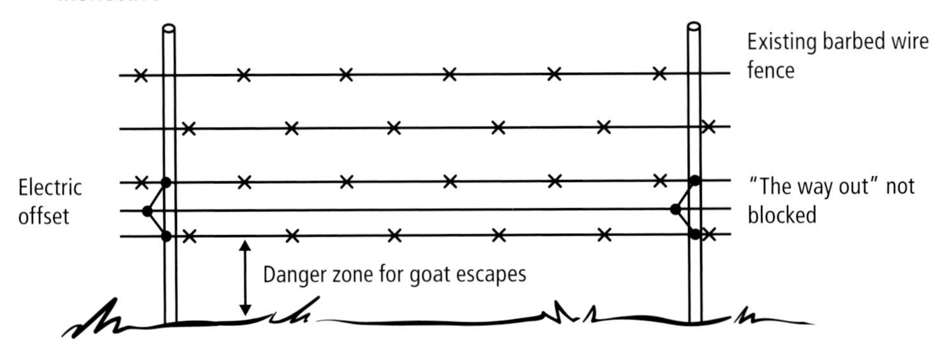

Effective

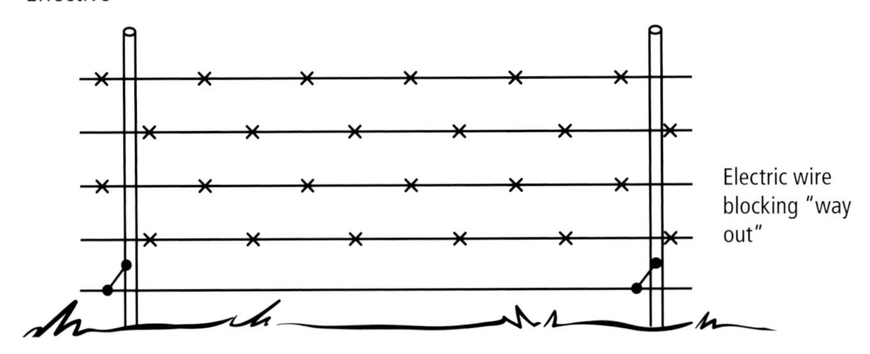

Boundary fences for this ranch in Texas, where deer need to be kept out of the goat pastures, are by necessity very tall (approximately eight feet). *Morgan Frederick for Three Mill Ranch*

For your fence to be effective for the vast majority of meat goats, your tester should show a reading of 5 lights (or 5,000 volts).

One downside to electric fences is that they work because goats aren't *willing* to go through the fence, not because they're *unable* to. For example, a goat might see a rose bush outside an electric fence and figure that the *cost* of eating the roses (getting shocked) is higher than the *benefit* of eating them (the delicious taste). But substitute a doe in heat for the rose bush, and have a buck in rut doing the cost-benefit analysis, and the result may well be different. The motivation for bucks in rut and does in heat is so strong that what works as a fence for protecting your garden won't be nearly as effective in keeping either sex away from the other when romance blooms.

Permanent Fence

Effective and Safe for Horned Goats

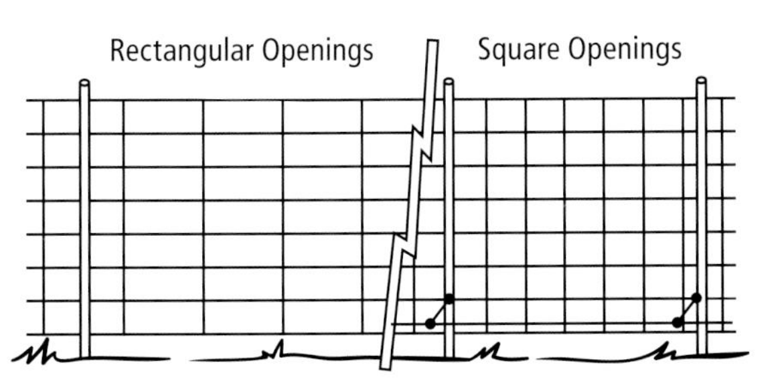

Electric "offset" will keep goats from getting their heads stuck

Electrified Netting

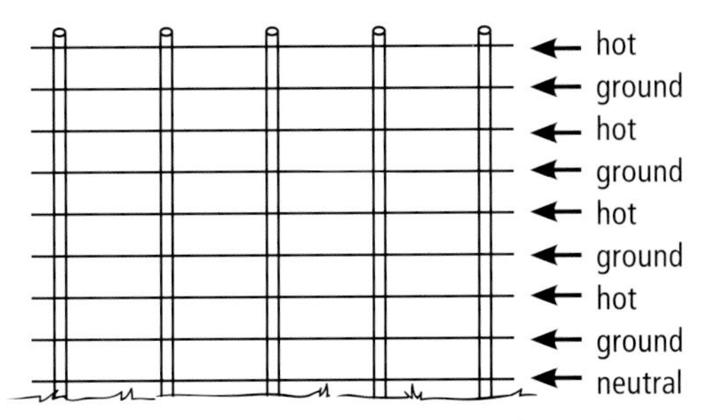

Portable, safe, visible 24/7 to humans and animals, virtually impenetrable by goats or predators; white PVC posts every 12 1/2 ft., plastic straps every 12 in.

4

ELECTRIC FENCING

For electric fences, the charger gets power either from a plug-in (regular house electricity) or from the sun via a solar panel that turns sunlight into electricity. Goat No. 1 (on right) is not touching the fence and thus enjoys a relaxing visit to the pasture, browsing and grazing. When Goat No. 2 touches the hot (electrified) wire, the electrical impulse comes off the wire, through the goat's muscles, through its legs into the soil, through the moisture in the soil back to the earth/ground electrode. What the animal feels is a momentary muscle contraction, such as a cramp, that is unpleasant but lasts for less than a second. Once an animal (a goat from inside the pasture or, for example, a coyote from outside the pasture) has experienced a shock, they tend to avoid any further contact with the fence. Goats that are raised with electric fence from a young age are typically much more respectful of electric fences than older animals who did not grow up with hot wire fences.

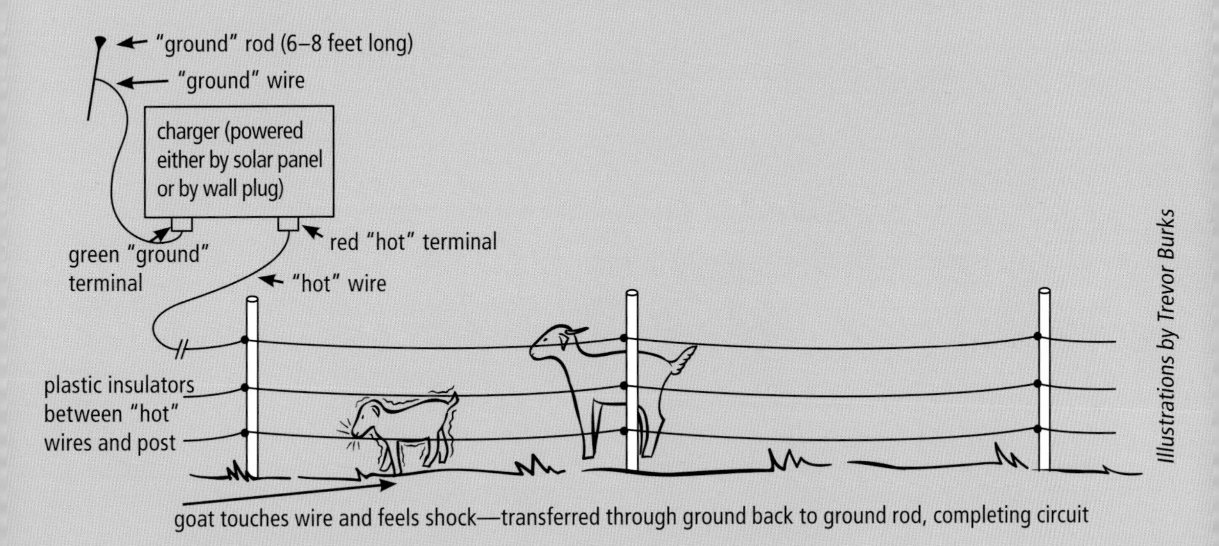

goat touches wire and feels shock—transferred through ground back to ground rod, completing circuit

Illustrations by Trevor Burks

The fences that subdivide the interior of this Texas ranch are approximately five feet tall but keep the Spanish goats in the pasture where they should be. *Morgan Frederick for Three Mill Ranch*

Meat goats learn very easily where the gate to their pasture is. These goats are in Pennsylvania. *Marcus Briggs, Dancing Heart Farm*

Distance and Staggered Schedules

When you're following a breeding program, it's important to keep male and female animals separated at certain times. But as you'll soon learn, keeping females on one side of a fence and males on the other will almost certainly lead to trouble sooner or later, as both sexes usually think that it's time for breeding season well before you do. In those cases, putting a physical buffer zone between the fences containing the two sexes, even if it's just the width of a driveway, can prove very successful in keeping the peace. Putting bucks on an island in the middle of a river has also done the trick for some producers.

For producers who don't have access to a small island, there is another way to prevent unplanned pregnancies that has proven successful: staggered

When you're following a breeding program, it's important to keep male and female animals separated at certain times.

grazing schedules. If you have a secure pen for each sex, the bucks can be penned during the day while the does go out and graze. Then, in the evening, the does can be put away and the bucks let out.

Many producers who use staggered grazing schedules give the daylight feeding hours to the does, as the nutrition they take in gives a more immediately visible result (quantity and quality of kids born and raised per doe) than it does in the bucks' production. Does don't need more feed than bucks, but they do need to take in enough to get pregnant, stay pregnant, and raise healthy kids. Letting does eat during the day, when it's easier to spot their food, will ensure that they receive adequate nutrition.

Both groups should get the hang of a staggered schedule within a few days. It will help if you can be fairly consistent about what time each group goes out and comes in, as goats adapt to routines very readily and seem to cherish schedules. If you happen to arrive late one morning, don't be surprised if the goats are standing at the ready giving you a look as if to say, "Humpf, *some* of us can tell time."

USING THE FACILITIES: HOW TO HANDLE YOUR GOATS

You may be able to walk up to any goat in your herd, slip a lead on it, perform the task that you wish to, let it go, and remember which goats you've already handled and which ones still need attention. If so, congratulations—I'm envious. Not so much because you can handle all of your animals, but because you can remember which ones are "done" and which ones are still "to do!"

If you're like me and have an imperfect memory, or you have goats that don't come when they're called, don't despair. Luckily for us some wonderful person thought up what is commonly referred to as a handling facility.

Handling Facilities

Handling facilities can be very basic—a simple pen with an extra panel for catching the goats and a way to let them out when the handling is over can be very effective. Handling facilities can also be more complex, with an alley and a weighing scale, different holding pens, and a loading ramp for easy transport.

This alley through a Montana hayfield is fenced with three strands of IntelliTwine. All three strands are hot (electrified) and keep kids and does out of the hayfield on the way to their pasture.

A pen that allows you to put all of the goats inside, perform the chosen task, and let the goats out will make life easier for both goats and people. *Tangy and Matthew Bates, Blue Creek Livestock*

Having a good handling facility does, in fact, make life easier when a task is just a task, and not a very difficult undertaking. Following are just a few of the many routine activities you might perform in a handling facility:

- Feel backstraps (this is a way of judging a goat's physical condition).
- Perform an eye check, to see if any of the goats are anemic.
- Trim hooves (although some producers use boulders or cement steps for their goats to use or play on and thereby trim their own hooves).
- Separate does into different breeding groups so that the buck of your choice sires the next generation of offspring, or separate unneutered male kids from the female goats to prevent the **bucklings** from proving that they can be fertile from approximately ninety days of age onward.
- Weigh the goats for your records or for planning your meat goat marketing.

As you can see, with the number of goat-related chores you'll need to do, it pays to have a functional, organized handling facility.

TIPS FOR A GOAT-FRIENDLY HANDLING FACILITY

BY JUSTA GOAT

- Keep in mind that whether we're part of a herd of ten or a thousand, we goats like to stay together and we don't like getting hurt. We also remember things that happen to us, good and bad.
- Please make sure that there aren't sharp things sticking out of walls or fences. These can catch and rip our skin, and that's scary, and it hurts.
- If the ground is relatively smooth and relatively dry, we'll find it much easier to get around from place to place.
- If the goats that have already been handled are let out into the pasture before I am, it makes me jealous. Why am I still stuck in the handling pen while they get to go eat? It's better to keep us all together if you can, or keep the goats that are already "done" in a pen near the handling facility.
- When things are scary, like if we have to go into deep shadows, we need to stop and look for a moment, if you don't mind. Apparently *you* can see into the shadows, but we can't.
- Please put the gates in corners of fences, have them open away from where we are toward where you want us to go, or roll out of the way. Closing any kind of gate too quickly so that it shuts on some part of any of us (even if it's old what's-her-name) hurts and is scary.
- Rounded corners (please) will make it easy for us to move into the alley for you.
- If you're going to use human helpers, please explain to them that slower is faster. Tell them not to yell and holler, not to use a **hot shot** on us, and not to pull our tails up over our backs to force us to move—that hurts *a lot*, which you would understand if you had a tail.
- If you use a herding dog, make sure that the dog actually stops threatening us when you tell the dog "at'll do" or "down." Remember, a good herding dog doesn't have to bite us to make us understand—all it takes is the look in his eyes. Trust me! (By the way, even though your eyes aren't quite as intense as those of a herding dog, if you stare straight at us, we'll stop. Just don't tell the other goats you heard it from me.)
- Keep things simple, smooth, and slow, and all of the work will actually get done faster. The writing-lady used to always try to do way too much at once when she handled us, but I guess she's finally learning to do just what's really necessary for our well-being, rather than doing everything possible.

The act of weighing goats deserves a bit more attention here. Weight is a nice number to know if we're talking about meat goats, and if you want to know a goat's weight, an accurate scale is invaluable. For example, imagine if you had market kids ready to sell but could only guess at their weights. In a situation like that, a meat buyer would be only too happy to tell you what the kids weigh! Hopefully, they wouldn't try to actively cheat you, but their job is to buy the pounds of goats you're selling at the lowest possible price. *Your* job, on the other hand, is to sell for the highest possible amount of money (the objective being to put your own children through college, *not* the meat buyer's). If you're going to get the best possible price for your goats, you'll need to know their weights.

Curved corners in the handling facilities prevent the goats from being bruised, and the goats therefore go into the pen and alley readily. *Dawn Van Keulen, VK Ventures*

These Boer goats in Minnesota move into the turntable freely. *Dawn Van Keulen, VK Ventures*

A turntable makes it simple and painless for the person handling the goats to trim feet or check the teat structure. *Dawn Van Keulen, VK Ventures*

A repurposed hog scale in Indiana gives an accurate measurement of a doeling's weight. The scale is portable, easy for goats to get in and out of, and never gets tired. *Paula Alley for Stoney Ridge Farm*

This ranch manager uses herding dogs to handle the meat goats. The dogs are invaluable employees, doing their jobs well and happily. *Morgan Frederick for Three Mill Ranch*

Getting Around

There are different ways to move meat goats from one place to another. You can lead the herd by luring the lead goat with treats; as that goat follows you, all the other goats will follow along, prompted by curiosity and the herd instinct that makes meat goats want to stay together. Usually, once you have a few goats following you, you'll have them all.

A four-wheeler is a very handy device for transporting goats over short distances in a dog crate or goat tote strapped to a cart that the four-wheeler pulls. Ours gets used dozens of times each day, both as a way to get from one place to another, and as a strong yet affordable version of a tractor. We also use ours to pull carts loaded with hay bales or to return a formerly sick goat to the herd after it has spent time in the TLC pen (TLC being short for tender loving care).

Usually, once you have a few goats following you, you'll have them all.

Some people use herding dogs to gather their goats, keep the goats together, and move the herd around. A well-trained working dog under a capable handler's control is an absolute joy to watch. The dog uses "eye," never its teeth, to make the goats stay, turn, and move in the direction and at the speed requested by the handler. "Eye" is how herding dog handlers describe the way a herding dog communicates with the goats and makes it clear to them what to do next. It is a language made up of a combination of eye contact, body posture, and movements that are well understood by the goats being herded. A good dog can speak this language of herding very quietly and gently, which results in the animals being herded doing as asked without fear, stress, or trauma.

The intent gaze of the herding dog is all that is needed for holding these Spanish bucks. The goats watch the dog closely, waiting for the next nonverbal communication. *Morgan Frederick for Three Mill Ranch*

At the handler's signal, the dog moves toward the goats to turn the bucks toward the chosen destination. Seeing a good herding dog work meat goats is like watching a well-choreographed ballet. *Morgan Frederick for Three Mill Ranch*

The goats are trying to snitch some dog food. When the guardian dog looks at them, the goats freeze in place.

HELP A GOAT: PLANT A TREE

What's so great about trees? Pretty much everything if you're a meat goat. Trees can give goats shelter from rainstorms and shade on sunny days; can provide small limbs, bark, and leaves to eat; and can be terrific toys for goat kids (and goats that are young at heart). Trees are also another crop that you can harvest for income from your land.

Meat goats use toys like this ramp to play on, which gives them good exercise and the goat owner an opportunity to laugh at the goats' antics. *Beth Sladky for Hillcreek Farm*

GIMME SHELTER FROM THE STORM

Over the years, several people have asked us why meat goats need shelter while cows don't. The simplest explanation is that a goat is built differently than a cow. A healthy, well-fed cow will have a sizeable layer of fat between its skin and muscles, which acts like a layer of insulation against the cold and rain. A healthy, well-fed

These Spanish bucklings show no evidence of weaning stress as they play on their spools. Covering the hole in each end of the spool prevents goat feet from accidentally getting caught. *Freddie Brinson, Brinson Pineywoods Cattle and Spanish Goats*

LET IT SNOW, LET IT SNOW, LET IT SNOW!

Despite their aversion to cold rain, all the meat goats that we've seen or heard about, living in places as different as Montana, Maine, Canada, and Pennsylvania, have no problem whatsoever with snow. And that's not only the Cashmere goats, who carry their own insulation with them in the cold months, but also Boers and Savannas. Those last two breeds are both originally from South Africa, but even goats recently imported to North America usually adapt within a season (one year at the longest) and can soon be seen playing in the snow with the rest of the herd.

meat goat will deposit excess calories as fat around its internal organs, at the core of its body, but not much under the skin. So a goat in a cold rainstorm feels like a person in the same cold rainstorm without a jacket on. The person's clothing would get soaked very quickly, and in very short order he or she would feel cold and miserable, just like the goat does.

What about warm rainstorms in the summer? From watching our goat herd, when the air tem-

Trees, brush, windbreaks, and sheds give goats the opportunity to get out of weather that is unpleasant to them.

perature and rain is warm, apparently the "Run for it!" reflex isn't triggered, and most goats will stay out in the rain, eating. But in the spring and fall, when the air and rain temperatures are cool, the first few drops of a rainstorm will send an entire herd at a run into the trees, under a lean-to, or into any available shelter.

Standing on his cement steps in Pennsylvania, this Boer-Spanish buck is sure that he is king of his world. His owner is very pleased that the buck is trimming his own feet. *Marcus Briggs, Dancing Heart Farm*

The owner of these lovely Pygmies doesn't groom her goats to make them look so well cared for; the goats do it themselves on their repurposed street-sweeping brush. *Staci McStotts, Critics Choice Pygmy Goats*

In the heat of a Wisconsin summer, this Kiko buck finds that his horn works as a pillow. He and the guardian dog share some shade. *Lynne Shonyo, Shonyo Farm*

Trees, brush, windbreaks, and sheds give goats the opportunity to get out of weather that is unpleasant to them. As I have urged before, let the goats choose to use shelter or stay out. Trust me, they know when they want to be in. We use a sixteen-foot-long, eight-foot-deep, four-foot-tall shelter that is open on the long side to the south because we have no trees. And why open to the south? For letting as much sunshine in as possible, and because our coldest winds come from the north. And why open on a long side? So that an assertive doe can't block the entire opening, the way she could if the narrow end of the shed were the doorway. This shed will give approximately 4,500 pounds of meat goats shelter from cold rain and cold wind. If the goats are approximately the same size or age, they tend to share the space quite well.

You've now gotten a good idea of the many ways to secure your goats, handle them easily and without fuss, and provide them with shelter. In the next chapter, we'll discuss how to keep your goats happy and healthy through proper nutrition.

DISASTER PREPAREDNESS

This is what nobody wants to think about. (I am grateful to Jon Kinsey of Paris, Ohio, for the information following).

Do you have access to a secondary source of water for your livestock (if the pump that makes the hoses run is electric)? Is it potable (drinkable) water? Is it gravity fed? In the winter (think blizzard), can you access it without it freezing over or freezing solid?

Do you have a secondary source of feed, be it hay or concentrates, that is readily available? Can you walk to go get it, with a cart or sled to bring the feed home?

Some items must have electricity to function (fencers, well pumps, milking machines, etc.) In these cases, is a backup generator or power source available?

Do you have a secondary source for fuel? In a widespread disaster, you may not be able to drive twenty miles to town to get fuel, or there may be no fuel deliveries in the town period.

Do you have someone trained and knowledgeable about your day-to-day operations? What would you do if you or your primary caretaker of the goat herd were to suddenly fall ill and cannot work for a week or more? Do you have someone who can step in and care for the animals?

Do you have ready access to cash? (Remember, ATM machines need electricity to work.) Do you have enough cash to buy your feeds, herd necessities, and take care of your family for a week or two if a natural disaster strikes?

Do you have an emergency plan for your family, including a preplanned meeting spot?

For larger operations that use large equipment daily, do you have access to a working secondary machine? Can what the machine does be done manually?

If you ever want to test your disaster plans, try living completely off the grid for a day or two (turn everything off, don't unplug your fridge and freezer, but tie a scarf through the handles so that you can't use them. No electricity, no phones, and only a battery-operated (or hand crank) radio . . . no TV either. You may just find yourself surprised at how many day-to-day things we do and items that we use that need electricity to function, and we take for granted they will always turn on.

Winter in Pennsylvania is no problem for a Cashmere goat who is wearing her own sweater and has nice green hay to eat. *Marcus Briggs, Dancing Heart Farm*

These heavily pregnant Boer does in Minnesota have rested and chewed their cud and are now coming back out to eat some more. *Dawn Van Keulen, VK Ventures*

On a damp, foggy Pennsylvania day, these goats enjoy soaking up the heat that their rock absorbed during yesterday's sunshine. *Marcus Briggs, Dancing Heart Farm*

As the doe returns from grazing to let her kids nurse, the kids are busy playing games in the shelter (it could be a game called "let's pretend we're pirates and walk the plank.") *Marcus Briggs, Dancing Heart Farm*

In Wisconsin, this doe and her kid seem to be curious as to why the photographer is inside the goat shelter. *Gayle Cayemberg for Hillcreek Farm*

CHAPTER 5

FEED, WATER, AND MINERALS

A meat goat's job is to convert food, water, and minerals into goat meat, either in his own body or in her kids. Your job is to give the goats access to what they need to thrive: food, water, and minerals.

Meat goats differ from other grazing animals such as sheep and cattle. Cattle and sheep don't have as much room inside for feed (as a percentage of their total body weight), so they must harvest or be fed a smaller quantity and higher-quality diet than meat goats. Meat goats, on the other hand, can consume a much higher percent of their body weight in feed. They can thus survive and thrive by eating larger *quantities* of lower-*quality* foodstuffs.

When a ruminant such as your goat chews its cud, the saliva that it secretes begins digesting the food.

In southern Indiana, this young Boer kid understands that summer is a great time to eat nice tall grasses. *Liberty Pastures*

CHEWING THINGS OVER

Meat goats, cows, and other grazing animals are known as **ruminants**. A ruminant is an animal that chews its cud, which is when it burps up or regurgitates a mouthful of already swallowed food, chews it again, and swallows it again. When a ruminant such as your goat chews its cud, the saliva that it secretes begins digesting the food. It also contains a natural antacid, which buffers the **rumen** (the first compartment of the stomach) and helps the goat fully digest its food, allowing it to get all the value possible out of it. This in turn enables the goat to eat more food, and thereby get more fuel out of its food in a benevolent cycle.

Boer goats in Kentucky head out to pasture. The next plant always looks better to a meat goat. *Dr. Ken Andries, Kentucky State University Land Grant Program*

People often say that when goats are happy, they chew their cud. They're right—an unhappy goat doesn't chew its cud, so when your goats are at rest (not sleeping, just resting), well over half should be chewing their cud.

When pastures are green, if there is a diversity of plant species and plenty of plant growth available, meat goats can be well-nourished by just eating from their pasture, regardless of whether the goat is growing, pregnant, nursing, or simply enjoying life (this assumes that the goat isn't struggling with a parasite infestation). There are many ways to make your available land more productive for your animals, including rotational grazing, irrigation, and planting more productive forages. But the challenge comes when

- there's not enough growth in the pasture for all of the animals, no matter how well you manage your land (or when nature covers the pasture with a blanket of snow);
- the goats should be growing (for example, to target market weight);
- the goats are heavily pregnant; or
- the goats are making milk for new kids.

These situations require goats to have *more* food available to eat, not less. Planning ahead to ensure that your goats have plenty to eat year-round is therefore very important.

FEEDING YOUR GOATS

Is there a recipe for success in feeding your goats? How much protein, energy, and carbohydrates do goats need to survive and thrive? Broadly speaking, meat goats need to consume about 3.5 percent of their body weight (on a **dry matter basis**); this amount should be about 13 percent protein.

The next branch looks better to meat goats, too, and the goats will even use one of their front legs to pull a branch lower. *Dawn Van Keulen, VK Ventures*

And what is dry matter basis? That is what you would have left if you took the goat's feed, wrung out all of the liquid, and weighed what was left. What's growing out in the pasture is probably about 20 percent dry matter/80 percent liquid, while hay is about 90 percent dry matter/10 percent liquid. This is why, if you were feeding your goats only fresh green plants, you'd have to harvest a *big* pile of feed—80 percent of what you would be feeding would be water.

Forage Feeding

Many meat goat producers follow forage-only feeding plans. This means that they feed their goats only things that are growing in fields or that have been grown in fields harvested and stored (excluding grains, which are often called **concentrates**). In the summer, their goats will harvest their own food in pastures or on weed projects. In the winter, if they live in an area where foraging is possible year-round, these producers keep their goats grazing or feed them hay.

Other producers follow a forage-based feeding plan. Their goats mainly eat from pastures or are fed hay, but the producers also supplement the goats' diet with grain. Some producers even mix and grind their own feed to give their goats optimal nutrition. (See the Resources section for information on getting a **Pearson square** to help you craft your own feed mix.) Producers can purchase commercially available sacked feeds that are specially designed for goats, but as with many things in life, the more convenient something is, the more it usually costs for the person using it and benefiting from the convenience.

No matter which feeding plan you choose to follow with your goats, make sure to refer to the information on **body condition scoring** that is described in Chapter 10. It's a handy way to grade your animals' physical condition. Neither extreme—be it overfeeding your goats until they're too fat to walk, or underfeeding them until they're too thin to stand—is helpful to achieving your goals. Overfeeding will waste your money and cause health problems in your animals. Underfeeding also causes health problems (particularly with reproduction), thereby hurting your goat's breeding potential and by extension your bottom line.

Pasture Feeding

If given a choice, goats prefer to eat forages that are chest-high or higher (measured on their bodies), and they will almost always eat top down, nipping off the tops of grasses and weeds before eating lower on the plant. This isn't just a behavioral oddity. It's actually based on the fact that goats are experts at browsing (eating leaves and taller forages, which grow well above the ground). In fact, goats' digestive systems and bodies are set up for browsing.

This affinity for browsing does have one unfortunate side effect: goats are susceptible to internal parasites. Often simply called "worms," parasites are tiny organisms that use the goat's digestive tract as their source of food, robbing the goat of vital nutrients. Parasites hatch from eggs and wait in dewdrops on the stems of forages; they're not usually found much higher than the bottom three inches of a plant, which means that a goat eating forage that is less than three inches tall is more likely to "get worms" than a goat that is eating taller plants. Once the parasites enter the goat's digestive system, they quickly reproduce and take as much nutrition out of the food going through the goat as possible. With such an infestation, the goat's body will be operating under a strong handicap: parasite load. It's hard for a goat to thrive or raise kids when parasites are stealing much of the goat's sustenance as soon as it's been eaten.

But wait, you say, I'll just use a chemical dewormer on the goat. Bingo, problem solved, right? Well yes, for a few days you won't have a problem. But if the goat goes back to the same pasture, with its short forage and parasite larvae, the whole cycle will start all over again. Besides which, many parasites are starting to

It's hard for a goat to thrive or raise kids when parasites are stealing much of the goat's sustenance as soon as it's been eaten.

show resistance to dewormer chemicals. So short of making the dewormer companies rich by treating your goats around the clock, what can you do if you want to follow a pasture-feeding program without running the risk of infestation?

Based on the lifecycle of the barberpole worm, which is probably the parasite that's currently causing the most trouble in the meat goat industry, there is actually a fairly simple solution (please note that I didn't use the word "easy"). Divide your land up into seven pastures. Graze one pasture for no more than ten days, then move to the next pasture. Cycle through all seven pastures and you should return to the first one sixty days later. If you're worried that a seventh of your land won't produce enough forage for all of your goats for ten days, you have a choice: add some land, subtract some goats, or purchase stored feed (hay). You don't have to buy the land—you can rent pasture, or, because goats are so good at eating weeds, you could actually get paid to have your goats harvest forages on someone else's land. This would be an example of prescriptive grazing. And remember, just because the goats devour the weeds once doesn't mean that weeds won't come back. Weeds are actually a symptom of the land being out of balance in terms of fertility and the species of animals that have traditionally been allowed to graze (please see the Resources section for places to get more detailed information on prescriptive grazing).

Of course you can't (usually) just turn your goats loose on someone else's land, even if they

Do meat goats eat burdock? With gusto! *Kim Hunter, Fossil Ridge Farm*

have invited you to. You'll need to check on a number of factors first: Are their fences goat-proof? Is there water for the goats to drink? Is there any shelter, or at least a windbreak that the goats could use to get out of bad weather? Can your guardian animal go along with the herd on the weed project? Ask what the landowners want to see happen, what they *don't* want to see happen, and how long they expect the goats to take to do the job, this time and on future visits. If you talk about things in advance, there should be few or no surprises.

Hay Feeding

If you're going to use hay as the basis of your goats' feed, or as winter feed, it's important to make sure that you're purchasing a quality product.

ROUGHAGE AND WHY GOATS LIKE IT

Many meat goat producers have watched in amazement as their goats walk right past the beautiful, dark green hay available to them in the fall and winter and start happily devouring dry, brown grasses or weeds. There actually is a scientific reason that goats prefer dry forage, and it has to do with slowly digesting carbohydrates (something called lignin) and dietary fiber. Wait! Before your eyes glaze over, there is an easier way to understand why goats eat what they eat. When I first acquired goats, my mentors told me that goats are "roughage busters." Roughage is much easier to remember than slowly digesting fiber or lignin, isn't it? You will see your goats actively seeking out roughage as part of their diet when adult goats leave the beautiful grass in the pasture to munch twigs or weeds, or a kid, just a few days old, nibbles on a piece of straw or dried grass as it starts eating solid foods, and thereby gives its rumen something to start developing digestive enzymes for. During a cold Montana winter, our goats will happily eat straw in addition to grass-alfalfa hay because as their bodies digest the roughage in straw, they get lots of body heat from the digestion. Alfalfa hay has much more protein in it than straw, but alfalfa is digested much more quickly by the goat and does not yield as much warmth (body heat) for as long a time for the goat. It's like throwing paper on a fire, rather than a log.

This Pineywoods heifer seems a bit perplexed to find goats in her hay feeder. *Freddie Brinson, Brinson Pineywoods Cattle and Spanish Goats*

One way to check hay quality is to ask the hay vendor for a copy of the bale's forage test, which determines nutritional values. Unfortunately, while a forage test can give you very useful information, it can also seem like it's written in a foreign language. A simpler method is to inspect the hay yourself.

If the hay is sold in what are called small square bales (typically between fifty-five and eighty pounds), buy a single bale. If the vendor only sells hay in what are called large round bales (a thousand pounds each or more), ask if he or she has a bale open for use and would allow you to take some home for the goats to taste test. Bring the hay home, cut the strings open if necessary, and check it out. Does the hay smell good to you even though you're not a goat? Do the flakes separate easily (a flake is one of the slices of hay that the baling machine packs together to make a bale), as well as the blades of grass within each flake, or is there mold in the hay, which will smell sour and cause the hay to clump together?

If the hay passes your inspection, it's time to ask the bosses. If you offer a few handfuls from a flake near the center of the bale to your goats, how do they react? Remember, your goats know hay, and the chances are fairly slim that they'll refuse to eat it outright. But does it seem like they like the hay a lot (quick sniff, big gobbling bites, and the hay is gone) or do they sniff and nose at it before gingerly tasting? Watching your goats carefully for their reactions will often tell you far more about a bale of hay than an official forage test can.

In a perfect world, hay would

- come from hayfields with well-balanced soils that have all of the minerals meat goats need and a diversity of plant species (sometimes called a pasture salad);
- be cut when just reaching maturity;
- be dried to the perfect moisture content; and
- be baled without ever getting rained on.

I probably don't need to tell you that it's rarely a perfect world, but that doesn't mean you shouldn't have some ideals in mind when you're buying hay for your goats.

Hay Feeders

One of the main complaints many producers have about meat goats is that they waste feed. When you raise goats, you'll see firsthand how picky they can be when sorting through hay, eating only what they want and ignoring the rest. Adding insult to injury, many goats will lie down on excess feed if they can, and when rising they will stretch, poop, and pee. Doing this to the hay ensures that no other goat will eat it as long as there's clean hay available. Since eating soiled hay risks contamination by internal parasites, the consuming goat will choose clean hay over soiled hay whenever possible. If clean hay

When goats have a place to put their front feet, they are not as likely to put their feet in their food. The colored stripes on the goats' hips are temporary marks indicating that the goats have received some treatment, such as a vaccination booster. *Dawn Van Keulen, VK Ventures*

This feeder allows goats to reach in and eat from a round bale and to drop their heads into the space between the wooden uprights. This way the goats can eat all of the hay without waste and without contamination. *Paula Alley for Stoney Ridge Farm*

isn't available, the goat will eat the dirty hay rather than starve, but the likely parasite contamination will keep them from performing at peak levels and make it more difficult for your meat goat enterprise to succeed.

To the rescue come hay feeders—farm accessories that keep feed hay safe and clean. The best hay feeders hold the hay a few feet off the ground, where the goats can eat the hay without stepping in their food (remember, goat feet have goat poop on them, and goat poop has worm larvae in it). A tray below

Watching your goats carefully for their reactions will often tell you far more about a bale of hay than an official forage test can.

When feeding grain or hay in a feeder, meat goats need approximately one linear foot of feeder space per 100 pounds of bodyweight. *Dawn Van Keulen, VK Ventures*

the feeder's hay rack will catch alfalfa leaves or small blades of grass as they fall, which happens whenever a goat pulls a mouthful of hay out of the feeder. A footrest gives goats a place to put their feet, which sounds odd until you watch a goat eating from a feeder—goats are very happy standing up on their back legs to eat, even if up is just a few inches high.

With goats lined up at a feeder, there should be approximately one linear foot per 100 pounds of goat. If there is less space per goat than that, the aggressive goats will get significantly more feed than the meeker goats. If you use concentrates as part of your feed, that tray below the hay rack will do a marvelous job of letting the goats eat their grain (or pellets). If the feeder is up against a fence, you'll be able to walk along the back of the feeder, reaching over and laying flakes of hay inside, pitchforking in hay from a round bale, or pouring in grain. Although, if you're pouring in grain, you might be running, not walking, so that your top doe doesn't eat it all!

Entire round bales also can be put into feeders. As long as there's enough space for *all* the goats to get at the hay (or there are multiple feeders), a round bale can take care of feeding a herd for a few days at a time. When the goats can eat on their own schedule like this, it's called ad lib or free choice feeding. The goats will absolutely love this feeding method, so if you want some very happy goats, make sure your business plan allows you the financial latitude to purchase lots of hay. If, on the other hand, you're trying to conserve hay, install your feeders in a place that allows you to control the goats' access to the food. Being able to close the goats out of the area where the feeders are will also allow you to put fresh bales into the feeders more safely, without the goats being able to get under the tires of your tractor, for example.

If you feed your goats hay from piles set out in a pasture, make sure that there is enough space between the piles so that no goat's bottom is positioned directly above another goat's food. Four to five feet between piles will ensure all the goats can eat their meal without the hay getting soiled. A nice side-benefit of feeding out in the pasture is that the small remnants of each day's feeding, and all of the goat feces already out in your pasture, will fertilize the ground without you ever picking up a shovel or pitchfork or starting a tractor.

After cornfields have been harvested, meat goats love being given the chance to devour remaining leaves and corn cobs. *Dawn Van Keulen, VK Ventures*

REUSE, RECYCLE, RE-FEED

The first time you see a goat munching on a thorny rose bush, you might wonder if goats are masochistic. The answer is no; they're just very good at harvesting food from plants with thorns because their lips are so good at extracting food from other, inedible material. This gives them a competitive edge over other species because goats can eat food that other ruminants often pass up.

You can put this evolutionary advantage to work for you by using goats to get rid of weeds on your land. For example, many meat goat producers in areas where wild roses (also called multiflora roses) are considered a weed really enjoy watching their goats devour the flowers.

You can also feed goats plant materials that you're no longer using. Old Christmas trees make a great goat treat. I don't remember who first told me that goats *love* eating evergreen trees, but our goats are very happy that they did. Now every year after Christmas, once the ornaments and tinsel have been taken off, used trees show up just inside the fence of our pastures. We and other local meat

Old Christmas trees make a great goat treat.

After the holidays, if undecorated Christmas trees are on the menu, the table manners of these Montana meat goats are nonexistent.

When goats are done with an evergreen tree, the needles are gone and much of the bark stripped.

goat producers also pick up old trees along Main Street in town. Cattle can't eat evergreens because doing so can cause spontaneous miscarriages. Goats, however, not only show no ill effects from eating evergreens, but also might actually experience fewer worm infestations as a result of their Yuletide feasting. As an added bonus, we find that eating evergreens gives a pleasant pine scent to the breeding bucks (even if it lasts for only a day or two).

Pumpkins, whether they've been carved or are extras from the pumpkin patch, are another post-holiday goat favorite, especially if the pumpkins are nice and ripe. Just make sure to remove any bulbs, candles, wires, or spikes before offering the pumpkins to the goats. To feed the goats, hold a pumpkin out at chest or shoulder height and let it drop to the ground. The resulting mess will be well enjoyed by your goats. Pumpkins are rich in vitamin A, which is good for the eyes, especially for night vision, and their seeds may have deworming value. Guardian dogs also tend to love pumpkin, to the apparent affront of most goats.

In the fall, your goat herd will also happily devour cornstalks and most other leftover plants from your garden. Don't feed your goats indiscriminately, however: Technically, potatoes are poisonous to goats. Our own goats won't eat potatoes even if they're the garden leftovers thrown into the goat pasture.

DRINK UP

The availability, cleanliness, and safety of your meat goats' access to water is very easy to overlook, but it's critical to their survival and health. Goats will drink approximately one gallon of water per hundred pounds of bodyweight per day. If the air temperature is high, they'll drink more, and when it's cold, less.

Without interference, algae can grow at the edges of standing trough water in just a few warm days. A few drops of Basic H in the water every few days should keep your troughs sparkling clean. Basic H is an organic, biodegradable surfactant that has no ill effects on the insides of a meat goat. Some meat goat producers even feel that adding

Availability of water is critical for meat goats. These kids in Kentucky apparently race each other for the top spot on this frost-free watering system. *Hebron Ranch*

Near Devil's Tower in Wyoming, these does and kids have been watered and are heading back out to pasture. *Carolina Noya, Flying D Cattle Company*

KEEP A RING AROUND YOUR ROSES

Here is a strong recommendation, based on personal experience: *Do not* let your goats mow the lawn if you have rose bushes growing in your yard, no matter how well you think you're protecting the bushes with temporary fences. I learned this lesson from a spry doe named Pelouche.

We had one beautiful yellow rose bush. To keep the rose safe from our goats, I carefully enclosed the bush with a circle of sturdy, bendable fencing. Minutes after I'd finished, the rose bush was mysteriously gone. In its place stood Pelouche, who'd somehow managed to jump inside the fence circle. She'd never jumped any barrier until that day, but the lure of the yellow rose was strong enough to give her wings.

Basic H to drinking water has a beneficial effect on worm loads in goats; others put whole black walnuts in their troughs for the same purpose.

Water should be easy and safe for the goats to get to, and possible for goats to get out of. If you use a large cattle trough for the goats, a set of cement blocks on the outside of the trough will give goats access to the water, and a set on the inside of the trough will allow a goat that is accidentally shoved in to get back out before it drowns. A more goat-friendly (and manager-friendly) way to water goats from a cattle trough is to put a fifty-gallon, twelve-inch-tall, goat-height trough right next to the cattle trough, and connect the troughs by way of a hose with a float on the end in the goat's trough. That way, as goats drink from their own trough and the water level drops, the float will allow more water to flow into the goat's trough. If cattle and horses also have access to the linked water troughs, a simple bar can prevent them from playing with the hose or (accidentally, I'm sure) pooping in the lower, goat-friendly trough.

The availability, cleanliness, and safety of your meat goats' access to water is very easy to overlook, but it's critical to their survival and health.

MIND YOUR MINERALS

In an ideal world, the forages your goats eat would provide them with all the minerals that they need to live. Unfortunately, this isn't always the case. That's why minerals are commercially available as supplements to your feeding program. Like vitamins for humans, minerals are nutrients that your goats need to build and maintain strong bodies, support normal growth and development, and help cells and organs do their jobs.

Despite minerals' health benefits, it can be easy to overlook them in your goat's diet.

Think about it: If you were to miss taking your vitamins one morning, or even a few mornings in a row, you probably wouldn't see any difference in your strength or vigor. It's pretty much the same with minerals for your goats, although you may well witness a food fight if you offer fresh

HOW TO MAKE A PVC MINERAL FEEDER

Here are the directions for making five 24-inch PVC mineral feeders for loose (granular) mineral. You can use either 3-inch or 4-inch size, but make sure all components are of the same size.

SUPPLIES

- One 10-foot length of PVC pipe (not thin wall)
- 5 wye hubs
- 5 caps (for tops)
- 5 test caps (for closing the bottom of wye hubs)
- PVC cleaner and cement
- Hacksaw for cutting the pipe into pieces
- Five lengths of medium-duty chain to go around top of feeder. For 3-inch PVC, use 15-inch chains; for 4-inch PVC, use 18-inch chains.
- Five lengths of medium-duty chain to go around base of feeder. For 3-inch PVC, use 30-inch chains; for 5-inch PVC, use 38-inch chains. (Wind it in a figure eight around the opening of the wye hub to hold the feeder up as well as to keep it from being lifted out of its cradle.)
- 10 screw-shut eyelets to connect the chains in a goat-proof manner

CONSTRUCTION

1. Cut the pipe into five pieces.
2. Glue the test caps into the bottom of the wye hubs.
3. Glue one wye hub onto one end of each section of pipe.
4. Put (don't glue it!) the cap onto the opposite end of each section of pipe from the wye hubs.
5. Let all the glue dry according to the instructions.
6. Hang the feeders on a fence near the goat's water source, facing away from your prevailing weather or under some shelter from rain.
7. The opening (where the goats eat the mineral from) should be at nose height for a grown doe when she stands near the feeder.
8. Put a cinder block (solid side up) near the base of the feeder for kids to use as a stepping stone.
9. Take the cap off the top, fill the feeder with granular mineral, replace the cap, show the goats where it is, and let 'em at it!

Occasionally you may need to stir the mineral in the bottom of the feeder to keep it loose because if goats with wet beards eat mineral, the mineral may cake or solidify after getting wet. A stick or the bottom spike from an accidentally broken tread in post works like a charm to stir the mineral.

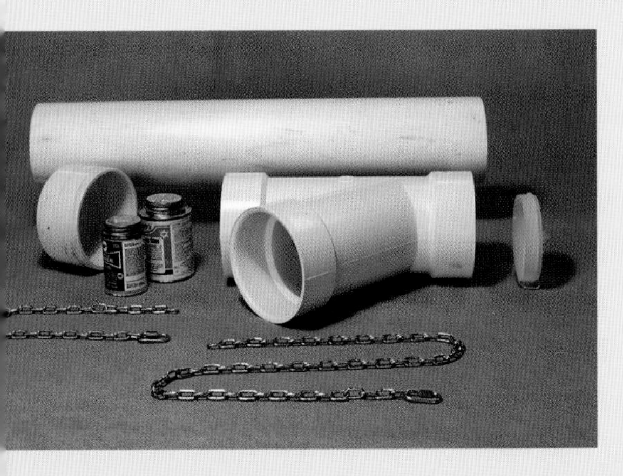

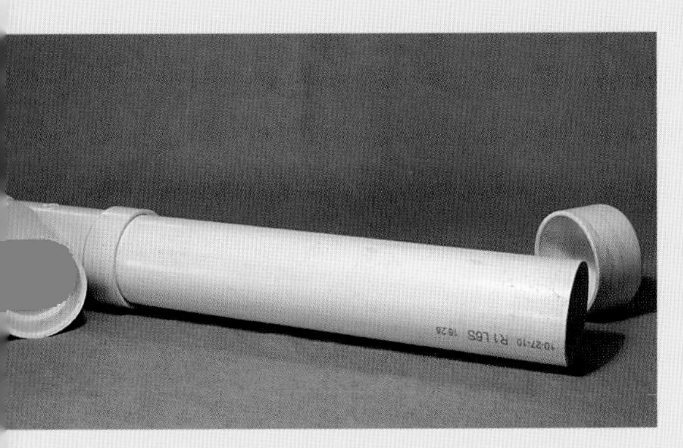

This Boer buck in Minnesota licks a mineral block to get trace elements that are not in the soil and thus will not be in the forage he grazes or in his winter hay. There are very few locations in North America that have perfectly mineralized soils. *Dawn Van Keulen, VK Ventures*

MINERAL CHELATION: WHAT IS IT AND WHY IS IT IMPORTANT?

Chelation is a process whereby minerals are protected so that they will actually get through your goat's digestive system and into the goat's circulatory system. It's not enough for meat goats to just consume minerals; they also have to be able to absorb them to receive any nutritional benefits.

Many things can block the availability of certain minerals. For example, our groundwater here in Montana has very high levels of iron (if the tap in a white sink drips, it leaves a reddish mark in the sink). Iron blocks the uptake of copper, which is a critical mineral for goats.

When we switched to a chelated mineral, it was clear that our goats were again getting enough copper. Solid-red goats that had unexpectedly grown white hairs on the edges of their ears and backs were once again solid red. Black goats that had been showing reddish highlights in their coats slowly returned to solid black. As the saying goes, seeing really was believing when it came to the benefits of chelated mineral.

If meat goat kids suddenly seem absolutely entranced by something in the dirt, sniffing intently and possibly even nibbling at the soil, there may be a certain mineral in that patch of earth that they are in need of. *Freddie Brinson, Brinson Pineywoods Cattle and Spanish Goats*

minerals to your herd after they've gone without for a few days.

Many meat goat producers are now offering loose (granulated) mineral to their goats by using a handy mineral feeder made out of inexpensive and readily available PVC pipe. The shopping list for such feeders and instructions for putting them together can be found on page 108. Mineral is also commercially available in a block. Both blocks and granulated mineral have pluses and minuses. Why not try out both and see what works best for your operation?

CHAPTER 6

HEALTHY GOATS

When raising meat goats, one of the many hats you'll wear is that of a chief physician. As a producer, you'll need to know when your goats are healthy, when they're not, and what to do in case of illness.

KNOW THE SIGNS

Before learning about how to tend sick goats, it's helpful to know what a healthy goat looks like. Goats are very curious creatures, and should be sociable (at least with other goats) and inquisitive about what's going on in the world around them. A healthy goat is interested in food and fresh water and will chew its cud after eating.

A healthy goat's eyes are bright and don't have any discharge, and the goat's nose is cool and dry. A healthy goat's coat is glossy and fairly slick, and when the goat walks it can cover the ground smoothly without limping or stumbling. The goat's fecal droppings should be firm, loose, round pellets, which is why it's easy to call them poo balls. The goat's urine should be clear (not cloudy or bloody) and a yellowish light brown.

A happy, healthy goat has its head and tail up, has bright eyes, and is interested in the world. A goat feeling poorly is listless, and its head and tail hang. *Illustrations by Trevor Burks*

A goat feeling poorly hangs its head and tail, is uninterested in its surroundings or other animals, and is largely indifferent.

A goat feeling poorly hangs its head and tail, is uninterested in its surroundings or other animals, and is largely indifferent. It doesn't want to eat or drink, which soon leads to hollows (that are not usually there) being visible just below the short ribs. The goat doesn't chew its cud. Its eyes are dull and may have discharge, and when you examine it more closely you notice that its inner eyelids are very pale. The goat's coat is dull and rough,

What color is a healthy goat? A shiny coat is indicative of good health in meat goats. *Pat Fuhr, Giant Stride Farm*

The Famacha card can help a meat goat producer identify whether a goat is anemic. This goat was rated D(4) and needed prompt attention. *Cori McKay for Smoke Ridge*

and it limps or stumbles when trying to walk. It doesn't have poo balls, but instead has diarrhea or clumped-together droppings. The goat's urine is likely cloudy or bloody. In general, it's A.D.R. (see page 115 for the sidebar Diagnosis A.D.R.).

When your goat is sick, you have two choices: You can take a "wait and see" approach and hope the goat gets better on its own, or you can address the A.D.R. right away. Since the chances of your goat magically regaining all its former vigor are pretty slim, you'll likely go for the second option. But what should you do?

You could call your mentor or veterinarian immediately, but both of them would probably appreciate it if you collected some specific details beforehand. So the first step when you're trying to treat a sick goat is to gather as much information about its symptoms as possible.

BEYOND THE PALE

Q: My sick goat has pale inner eyelids. What does this mean?

When a goat is anemic (meaning its blood has fewer red blood cells than it should, and therefore can't move oxygen to the cells of muscles and organs), the inner eyelid will be a pale pink, a very pale pink, or nearly white.

Given a goat with pale inner eyelids, most producers would assume that the goat was badly infested with parasites; this is not always right (usually, but not always). Copper deficiency can also cause anemia. Copper is one of the most important minerals that goats need in order to function, and iron in the groundwater can block uptake of copper in a goat. So even though anemia in humans is typically an iron deficiency, in meat goats an iron surplus can trigger anemia.

THE DOCTOR IS IN

Before you start checking the animal, get the goat into a small pen or stall where you can handle it. This is helpful because no matter how friendly your goat normally is, it might not be eager to have its temperature taken, for example. If that happens, you'll appreciate having the goat in a small, controllable space. (Try not to get frustrated with your sick goat's behavior—just remember how it feels when *you* have a bad cold.)

To give your goats a thorough check-up, it's essential to have a meat goat tool kit (see photo, page 34). A well-stocked kit might include the following:

- a lead rope
- a stethoscope
- a thermometer (with disposable covers if the thermometer is digital)
- a watch with a second hand, or a stopwatch
- a small notebook and pen
- a cheat sheet listing specs for a healthy goat (maybe written on the inside front cover of the notebook?)

Many producers have found that the money you spend on a vet, especially in the early years of learning about your animals, will pay itself back very quickly.

Using your tool kit, check and record the goat's temperature and heart rate. Don't forget to listen to the goat's breathing too—once you're within a few feet of the goat's face, you may well be able to hear labored breathing from congestion without even needing the stethoscope.

Now that you've noted the goat's vital signs, the next piece of the puzzle is to figure out what is wrong and what to do about it. One option is to call your mentor and explain what's going on with the goat. Your mentor might not be a practicing veterinarian, but he or she will probably still have lots of helpful advice.

If your mentor isn't available, call the vet. Ask the office manager whether the vet is willing to answer questions by phone and whether there's a charge for that (vets care very much about the animals in their care, but they do need to make a living). If there is a charge, strongly consider paying it if you can—many producers have found that the money you spend on a vet, especially in the early years of learning about your animals, will pay itself back very quickly.

Depending on what your vet says, you might have to give your goat medicine.

Something we keep tucked in the corner of our refrigerator is a tube of probiotics. Probiotic supplements usually come in a blue and white tube with a dial. When you turn the dial to the correct bodyweight, the tube dispenses the correct dosage of the contents. Probiotics contain the good bugs that a meat goat needs to properly break down food. Giving your goat probiotics will help its digestive system get back on track after an illness.

You might have another kind of goat medicine, not in your refrigerator, but on your land.

Willow trees, from which aspirin is derived, can be a wonderful pick-me-up for ailing goats.

DIAGNOSIS: A.D.R.

I learned about A.D.R. when I was first starting out with meat goats. An early mentor of mine, upon seeing one of my goats, said with worry in her voice, "Oh no, that goat is A.D.R." Since she was a practicing veterinarian, I fell for it and worriedly asked, "What's *that*?" With a completely straight face, my mentor answered, "Ain't Doing Right."

She might have been teasing me, but sometimes A.D.R. is actually a very good description of what is wrong with the goat. It might not be an official ailment, but diagnosing a case of A.D.R. can be the first step toward finding out what's really wrong with your goat.

HEALTHY GOAT SPECIFICATIONS

Body temperature: 101.5–103.5°F (take rectally)
Heart rate: 70–80 beats per minute (17–20 beats per 15 seconds on your watch)
Respiration at rest:
Adult: 12–15 breaths per minute (3–4 per 15 seconds on your watch)
Kid: 20–35 breaths per minute (5–9 per 15 seconds on your watch)

One meat goat producer we know gives sick goats willow tea and, although that producer does use commercial medications when necessary, finds that letting a goat eat some willow leaves or drink a cooled willow tea frequently takes care of many minor ailments.

Whatever course of treatment the vet prescribes, follow it carefully. Make sure any medicine dosages are given correctly, and always follow the full course—don't stop a few days early or skip dosages.

PREVENTING PARASITES

The most common health problem you're likely to come across is a parasite infestation. Parasites, which can cause a range of digestive and circulatory problems, seem to be one of the biggest challenges meat goat producers face. You've already learned how eating habits can help goats avoid parasites (Chapter 5), but what else can you do to prevent these tiny pests from troubling your herd?

The most common health problem you're likely to come across is a parasite infestation.

If you're living in the right area, you might be able to prevent parasites by simply letting your goats forage. Several forages out there contain **tannins**, which might not be a miracle remedy, but show great promise in retarding the ability of parasites to successfully reproduce.

Tannins are the chemical substances found in many plants. The word *tannin* comes from the old High German word for an oak or fir tree. (Remember singing "O Tannenbaum" for the holidays?) Oak tannin is used in the process of turning animal hides into leather. Tannins protect plants from various microorganisms and apparently can help protect your goats from internal parasites.

There is a plant known as *Sericea lespedeza* that is naturally high in tannins and will therefore suppress parasites. Goats will consume *Sericea lespedeza* with great gusto. In some parts of central and southern United States, *Sericea lespedeza* is considered a noxious weed, and people are trying to combat its growth. If you live in such an area, perhaps you could get your goats involved in a weed control project. You'd be helping others while your goats got worm-fighting medicine for free. Other plants that contain significant amounts of tannins include acacia (wattle), sainfoin, trefoil, mangrove, eucalyptus, myrtle, oak, maple, birch, willow, pine, and sorghum.

This Boer goat in North Carolina doesn't know that it's benefiting from the parasite-suppressing tannins in *Sericea lespedeza.* It just knows that the plant tastes great. *Jean-Marie Luginbuhl, North Carolina State University*

RUMEN-ATIONS

You already know that goat are ruminants, or grazing animals, but what exactly does that mean? What is a rumen, and what does it do for a goat? How does it function normally, and how do you get it back on track when it doesn't?

The goat actually has four stomachs, of which the rumen is the first. In the first month or so of a goat kid's life, the rumen is very small and the milk that the kid drinks from its mother passes directly to the final, or true, stomach. As the kid begins to mature, it starts to nibble on roughage, and the rumen begins to manufacture microorganisms that will break down the goat's food.

The rumen of a mature goat essentially acts like a fermentation vat, where good bugs (beneficial bacteria and fungi) produce the enzymes that break down fiber in the goat's feed. A happy goat, after harvesting or eating its roughage, will periodically burp up a cud. When this happens, the goat regurgitates its food, chews it again, and reswallows it. Each time the goat does this, not only are the food particles further broken down, but additional saliva is mixed with the food, which is one of the first steps in efficient digestion.

After the food particles are small enough, they pass to the goat's second stomach or **reticulum.** The reticulum operates like a sieve, filtering out any nonfood items that the goat accidentally swallowed. The third stomach, called the **omasum**, is where the water is removed from the fermenting bits of what used to be the goat's food, along with nutrients and the energy ingested by the goat as fatty acids. Once this stage is complete, the nearly digested cud passes into the **abomasum**, or true stomach, where stomach acids finish breaking it down, just like in our human stomachs. From the true stomach, the particles pass into the small intestine for absorption, then move into the large intestine. From there they finally emerge as small, round droppings.

The rumen can malfunction for a variety of reasons. If the goat is trapped lying on its side, and cannot get up onto its breastbone (or to its feet), the goat cannot burp up a cud. More importantly, it cannot burp up the gases that form in the rumen from the fermentation of the food and it will bloat and can die. **Bloat** can also occur if the goat consumes a large quantity of foods that it isn't used to eating and therefore doesn't have the correct microorganisms to break down and digest. In either case, the gases trapped in the rumen will continue to expand if unchecked, causing the rumen to swell against the lungs. The goat would eventually suffocate to death.

Bloat can also occur if the goat consumes a large quantity of foods that it isn't used to eating

Luckily it's possible to spot and treat bloat before it reaches fatal levels. One early indication is that the goat walks stiffly, with its rear legs further behind its body than is normal. Another sign is that the goat will be visibly lopsided, with it's left flank protruding higher than its spine.

To treat the bloat you'll need a large-gauge needle (such as a fourteen or twelve). Gently but firmly push the needle into the goat's distended side, high up. What you are attempting to do is let the air out of the bloated rumen. You should hear a whoosh, there will be a distinct smell, and the goat will look a great deal more normal. Remove and recap the needle. *Now* you can call your mentor or the vet for further assistance if necessary, as without immediate action, the goat would have been in excruciating pain and may have died.

Frothy bloat is a particularly unpleasant occurrence that is triggered when the animal gorges itself on lush, moist foods, such as alfalfa. Frothy bloat, unlike regular bloat, is not one bubble of air but thousands. Fortunately, like with regular bloat, the goat can be saved with *immediate* action.

WHAT'S SO GREAT ABOUT GOAT POOP?

Goat manure is great for gardening because it has a very high level of nitrogen, which plants thrive on. Only bat guano has more nitrogen than goat poop. Since few people have bat guano readily available for their garden, goat poop is a handy alternative.

Goat poop doesn't normally smell very strongly or attract flies, and the pellets are easy to transport to and use in gardens or flowerbeds. Most meat goat producers want all of their goat poop to add organic matter to their own soils, but bagged-up goat droppings could easily become an additional product to sell.

A rock might not seem like a comfortable spot to us, but this one-year-old Boer-cross doe and her daughter are enjoying its warmth. *Priscilla Ireys, Critton Creek Farm*

To treat frothy bloat, there's a liquid called Therabloat that you mix with water. You then use a needle-less syringe to push the fluid into the goat's mouth, making sure that the goat swallows. If you act quickly enough, as the froth is still forming, the Therabloat will cause all the bubbles in the froth to pop. The goat will belch (once or repeatedly) and will start to look visibly better. Recovering fully from frothy bloat is not as easy for the goat, so it may be noticeably subpar for a day or two.

NOBODY WANTS LOUSY GOATS!

If only one or two goats are scratching their backs, especially in spring when they shed their winter coat, it's a normal part of being goats. If many goats are scratching at the same time, or goats have small plucks of hair standing up on their backs, check the goats for lice. These are a parasite, such as fleas on a dog, that bites (not too surprisingly, called biting lice) at or sucks through the skin of a goat (sucking lice). Lice can easily be controlled with chemical delousers.

We find that lice, if they occur, will be on the smallest of a set of triplets or on an animal that for some reason is compromised. For us, lice are a warning sign that a goat may be at risk for other problems because the lice have gotten past the kid's defenses or ability to deal with them. Many goats can have some lice, and you will never even notice it, because it's a minor irritation rather than a major problem for that goat. When a louse lands on a slightly weaker goat, such as a triplet kid who might not be competing for the available milk as well as its siblings, it seems like the louse population can explode on that kid.

What you will notice when a kid is suffering from a louse infestation is the kid standing sort of hunched up, with all four feet very close to each other as the kid tries to minimize the surface area of its body in order to stay warm. If you can catch that kid (by hand or by using a nifty fish net) and you ruffle the hair on the top of the kid's head back, you may well see tiny reddish things moving around on the goat. Those are biting lice (yuk). A single dose of ivermectin pour-on (for cattle) at 1cc per 22 pounds of bodyweight will take care of the problem. A producer who chooses to use **homeopathic** medications may recommend tea tree oil.

IF YOU LOVE YOUR COFFEE, THANK A GOAT!

Legend has it that a goat herder discovered what coffee can do for people after he noticed how much more boisterous than usual his goats were acting after they nibbled the "berries" from a certain bush. He might have thrown some of the beans in boiling water and, as they say, the rest is history. It's a nice story and a wonderful beverage, and watching young goat kids play, bouncing with exhilaration and joy at just being alive, it's easy to believe the legend.

Dawn Van Keulen, VK Ventures

UH OH, IT'S AN OOCYTE (PRONOUNCED OH-OH-CYTE)

Coccidiosis is the disease of goats having an abnormally high number of coccidia, a naturally occurring (always around, but normally in limited numbers) protozoan (single-celled) parasite. Goats ingest the oocyst (egg) from the environment, where they must have air, moisture, and warm temperatures for development. Moist areas out of direct sunlight can harbor infective oocyst for a year or longer. Intestinal cells of infected animals are destroyed by the parasite, leading to poor animal performance. Symptoms of coccidiosis range from loss of appetite and slight, short-lived diarrhea to severe cases involving great amounts of dark and bloody diarrhea and, in some cases, death. The most susceptible goats are very young kids and kids that are being weaned. Adult goats can develop coccidiosis if they are stressed or if they are moved into an environment that is heavily infested with oocysts, especially in crowded conditions.

The smoothness and thickness of a goat's horns, like rings on a tree, can show you how well the goat grew each year.

CARING FOR HORNS

Horns are part of a goat's self-defense mechanism, in addition to speed, agility in climbing, and an ability to change direction very quickly. A buck defending his does from the attention of another buck will use the base of his horns to head butt the other buck. Goats also use their horns to scratch their own backs (wouldn't it be nice if humans could do that?).

The horn that you see when you're looking at a goat that hasn't been dehorned is actually an outer covering of keratin (like our fingernails) over an inner living bone. This bone grows a little bit each year, with the horns of intact male goats growing more than the horns of females.

The smoothness and thickness of a goat's horns, like rings on a tree, can show you how well the goat grew each year. If there is a visibly thinner part of both of a goat's horns, that animal was sick or hungry during that period, after which it recovered and horn growth got back on track. If the visibly thin spot is on only one horn, that spot is the goat's favorite place to rub.

Goat horns are virtually indestructible. That means that you will experience only a very small percentage of them breaking on your watch. If the part of the horn that breaks off is the part farther away from the goat's head and is only keratin (like our fingernails), the damage is only unsightly. If the living bone (with nerves and blood vessels) in the half of the horn closest to the goat's head breaks, it will cause the goat pain as well as profuse bleeding.

There are actually does underneath the winter pelts that are shedding off now that spring has arrived. If both undercoat *and* outer hair falls out, suspect a zinc deficiency. *Marcus Briggs, Dancing Heart Farm*

This magnificent Cashmere buck in Maine has a problem: What delicious plant should he eat next? *Black Locust Farm*

Using a blood stop powder or spraying the base of the broken horn with an antiseptic spray will prevent infection. You may choose to remove the broken horn completely rather than leaving it to dangle and irritate the goat (your tool kit could have a pair of snips or a utility knife in it).

Most meat goat producers find that if you give a goat a chance, it will be healthy and happy. "Head up, tail up, bright-eyed, and interested" describes a meat goat that is going about its business of being a goat. Goats that are A.D.R. might need to get their rumen (and the other three stomachs!) back to working properly so that they will feel better and start chewing their cud again. You should try offering the goat some willow, or plants with tannins in them, before reaching for something from the medicine chest.

CHAPTER 7

LIVESTOCK GUARDIANS

In the ongoing battle to protect their herd from predators, many meat goat producers use guardian animals such as dogs, llamas, or donkeys. Guardian animals protect the goats around the clock, sleeping, eating, and drinking with the herd.

Man has used guardian dogs for centuries in much of the world to keep livestock safe. However, they have become common in the United States only in the last few decades as other methods of predator control were banned by law and producers sought new ways to keep their goats safe.

What follows in this chapter is a list of frequently asked questions about guardian dogs: what they do, where to find them, and how to keep them as happy and healthy as your goats.

How do guardian dogs protect animals?

Guardian dogs are successful when they disrupt the stalk-chase-grab-kill behavior of predators (or the swoop-grab-fly-away-with-the-goat pattern of birds of prey), and all of the goats remain alive and uninjured. Guardian dogs also guard against predators by scent-marking the territory they are protecting with urine and feces. Predators (including coyotes, foxes, and wolves) understand these odors just as clearly as we understand what giant billboards or stop signs tell us.

Guardian dogs protect herds by using their keen senses to identify when something unusual and thus unwelcome is near their goats. A dog's sense of smell is said to be a thousand times as sensitive as that of humans, and their hearing can detect higher frequencies than a human can. Because dogs have the ability to position their ears independently of each other, a dog also can pinpoint the origin of a sound much more accurately than a human can. The guardian dog uses both its sense of smell and hearing to determine whether any creature is near its herd that doesn't belong there.

Guardian dogs protect herds by using their keen senses to identify when something unusual and thus unwelcome is near their goats.

If the guardian notices a trespasser, it will first bark a warning, then move toward the uninvited guest. The vocal warning will usually cause the unwanted visitor to leave, but if the departure is not speedy enough in the guardian dog's opinion, he or she will rush toward the trespasser in a threatening manner. Actual fights are extremely rare, as predators usually choose to run away unscathed.

Why use a livestock guardian dog?

Because livestock guardian dogs are very good at their jobs and because they're always on duty, even

This guardian dog in Texas may look like she's sleeping, but her sensitive ears and powerful sense of smell never go off duty. Lying on the moist ground around the pond lets her cool off. *Carol DeLobbe, Bon Joli Farm*

when it looks like they are fast asleep. Guardian dogs work around the clock, not recognizing weekends or holidays.

How do you make a dog guard a particular animal or herd?

An adult guardian dog instinctually wants to protect what it is raised with. The species of animal the dog will protect is the species that it scented consistently when it was growing from newborn to adolescent (between two and twelve weeks of age). If the puppy predominately smells and is around meat goats during this period, that is what it will want to protect when it matures. If it is raised with humans during the developmental period, it will be bonded to people and will guard them.

What kind of a dog is best suited to be a guardian dog?

Many breeds and composite-breed dogs guard meat goats from predation all across North America. Simply in alphabetical order, with no intention of conveying any ranking, you may hear or read about the following dog varieties: Akbash, Anatolian, Caucasian Ovcharka, Estrella mountain dog, Great Pyrenees, Kangal, Karabash, Karakachan, Komondor, Maremma, Pyrenean Mastiff, Sharplaninac, Sivas, Slovak Cuvak, Tatra, and Tibetan Mastiff. To a large extent, the specific breed of dog you use doesn't matter as much as its ability to guard your goats from predators.

Neither sex of guardian dog has been proven to be significantly more or less successful than the

Even a gentle slope gives this dog in Ohio a good vantage point from which to watch for disturbances in the goatherd. *JS Family Farm*

other. Whether dogs are neutered or intact also doesn't affect their ability to guard goats, with one possible exception: Intact dogs of either sex can (and probably will) be distracted from guarding by the urge to reproduce. This can leave the dog's herd unprotected while it's off searching for a mate or put the herd under even greater **predator pressure** if potential mates come to the guardian dog's property in search of romance.

In general, whatever the breed or sex, a good guardian dog will display the following:

- trustworthiness (doesn't injure its charges)
- attentiveness (pays attention to what's going on with its charges)
- protectiveness (cares about the well-being of its charges)

At what age do livestock guardian dogs become effective protectors? Do they work well from weaning age on?

No. Young dogs are still growing into their bodies and their jobs, like all young animals do as they mature. Most guardian dogs will exhibit protective behavior from a few months of age on, but typically won't mature into their role as flock guardians until approximately one year of age.

What is a livestock guardian dog worth, and how much will one cost?

Remember the worksheet in Chapter 3, where you calculated what your perfect meat goat was worth? If a livestock guardian dog saves just one goat kid per year for you, in a nearly decade-long working

Llamas are also used as guardian animals and can graze along with the goatherd. *Carolina Noya, Flying D Cattle Co.*

In West Virginia, the female Maremma stays near her goats as they enter the woods to eat some more delicious leaves. *Priscilla Ireys, Critton Creek Farm*

life, the dog can save your meat goat enterprise many hundreds, if not thousands, of dollars in market animals alone. What if half of the kids saved by the dog are female goats that you would have put into production? How much more would the dog be worth to you then? (That could be an interesting calculation, but to keep things simple, let's just say even *more* money.)

In terms of price, a guardian dog can range from under one hundred to over one thousand dollars, depending on the quality and age of the dog. Pups will typically cost less than older, working dogs, as they still need time to mature into their job. Some pups will need extra attention from the herd manager to correct juvenile delinquent behavior, which is a normal phase of growing up but does need to be stopped rather than ignored.

Very rarely, a mature, working livestock guardian dog will become available for sale. It will probably be much more expensive than a pup but would offer new guardian dog owners all of the benefits of a good dog without the wait for the dog to grow up.

Where can you get a guardian dog?

Many meat goat producers who use guardian dogs will have pups available occasionally. They're the best source for a good working dog that has been raised properly and will therefore do what you want it to. You can also ask friends and mentors for leads, or look at the advertisements in your favorite meat goat magazine.

Does a livestock guardian dog need special training to do its job?

No. Guarding behavior comes from how the dog was raised and its instinct to keep its herd

Many meat goat producers who use guardian dogs will have pups available occasionally.

undisturbed. Young guardian dogs may need some management, as adolescent (less than one year of age) guardian dogs occasionally decide that it would be fun to chase goats, especially when goat kids run. Chasing can lead to pulling on goats' ears or tails, which is inappropriate behavior for a guardian dog.

Some pups never exhibit bad behavior, others need only a single correction to mend their ways, and still others need to be caught in the act and yelled at multiple times. Your strongly disapproving tone of voice will be much more effective than any physical correction could be.

When it comes to on-the-job training, you'll likely have an ally in goats: A mother goat will often deliver a very clear, corrective head butt to a wayward pup that's teasing her kids. A pup that needs a time out can be "put in jail," physically apart from but still able to smell and see the goats for a few days, or it can be put in with the big bucks. The bucks will put a quick stop to inappropriate puppy behavior.

This livestock guardian is a Karakachan, a brindled dog originating in Bulgaria. As long as his goats are climbing on trees undisturbed, life is good. *Priscilla Ireys, Critton Creek Farm*

Guardian dogs like to have a good view of the herd, frequently picking a high point from which to watch their goats.

How much should I bond with my guardian dog?

You may have heard that one should never touch, pet, or make eye contact with a guardian dog. You may even have heard that any emotional connection to the dog whatsoever is a bad idea. With all due respect to people who disseminated those ideas, in the mistaken belief that any human-dog interaction would reduce or block the dog's

While its mother has quickly gone to get a drink, this newborn goat kid takes advantage of the guardian llama's shadow. *Weed Goats 2000*

attention to the livestock, this notion couldn't be more wrong. We now know that the bonding period between two and twelve weeks of age sets which animals the dog will guard, and having a human in or near the herd for a few moments per day will not make the dog forget its job.

As longtime and very satisfied guardian dog users, we know that handling pups as they grow gives you the ability to handle the mature dog without having to resort to tranquilizer darts or drugs, and it makes life much more pleasant for you and the dog in many instances. Being able to call the dog to you and slip a collar and leash on it so that annual vaccinations can be given, or porcupine quills removed, is also a very good thing for you, the vet, and the dog. None of these things are likely to happen easily (or at all) if you don't have a strong connection to your dog.

When two guardian dogs share responsibility for one herd of goats, they frequently alternate between staying close to the goats and checking the pasture's perimeter. Here one guardian seems to check in with the other before heading out to patrol. *Kaylene Larson, KM Larson Ranch*

These kids in Idaho have been left by their mother in the care of the guardian dog while she goes to eat. One kid is already happily snuggled up against the dog's shoulder. *Weed Goats 2000*

What and how much do I feed a guardian dog?

Since few of us meat goat producers have the byproducts from cheesemaking (whey) to feed our guardian dogs as they do in Europe, we typically feed them dog food. The best dog food you could buy is the one that keeps your dog healthy and capable of doing its job well.

At present, most working guardian dogs in North America tend to mature as rather large dogs (approximately a hundred pounds), which means that they are slow-growing, late-maturing dogs compared to the typically smaller dogs that are kept as pets. Letting young guardian dogs' bodies grow at a moderate pace by feeding adult dog food starting at their introduction to solid foods will preclude the behavioral or health problems that many high-protein or puppy growth menu dog foods can lead to in early or later life.

How do I keep the goats out of the dog's food?

- Feed the guardian dog(s) at a consistent time each day.
- Feed the guardian dog(s) while the goats are busy with their own hay or fresh pasture.

"JUMP PANEL" DIMENSIONS

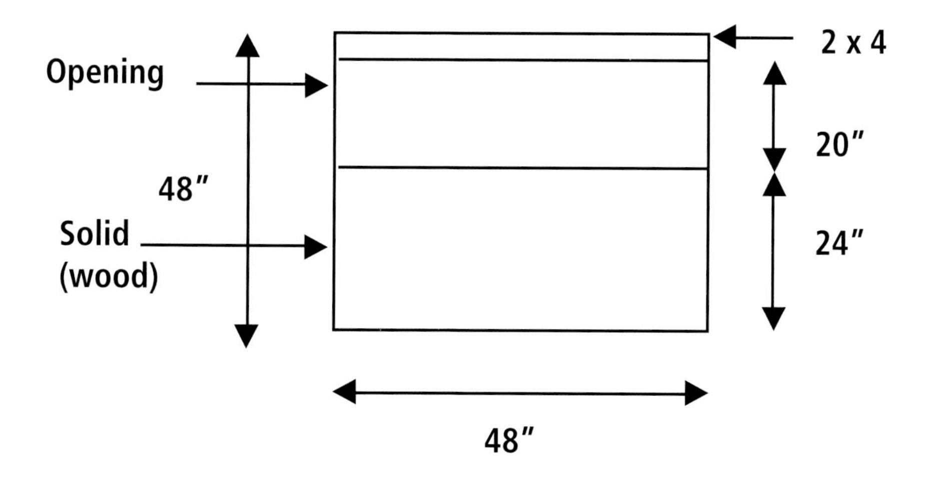

- Use treats on the dog food as needed to get the dog to finish all its food. This way, there won't be leftovers for the goats to snitch.
- Don't overfeed the dog(s), again so that they clean up their meal rather than leaving leftovers for wily goats.
- Use a jump panel for the dog to get into an enclosure where a self-feeder is located. The goats won't be able to get in, and the dog can eat in peace. Use a plastic rather than metal self-feeder to prevent heat or cold affecting the dog's ability to use the self-feeder comfortably year-round, and remember to check for rodents in the self-feeder periodically.

The guardian dog should be with the goats all the time, not up at the house.

How do I keep my dog with my goats?

A trustworthy, attentive, and protective guardian dog will stay with the herd by preference. If a guardian dog doesn't have reproductive organs playing their siren's song, it will be even more likely to keep its mind on its job.

On a beautiful fall day in Alberta, a Maremma is quiet because his goats are undisturbed. *Pat Fuhr, Giant Stride Farm*

The fence that holds the goats should also hold the guardian dog. A top offset hot wire should preclude any climbing over the fence by the guardian dog.

The guardian dog should be with the goats all the time, not up at the house.

Will a guardian dog be aggressive toward herding dogs?

Guardian dogs are typically herded right along with the goats by the herding dog. Herding dogs should be under good control by the handler and not use undue aggression on the goats. If the actions of the herding dog cause one of the goats to cry out (an unusual event in a normally calm and happy goat herd), the guardian dog is put in a difficult position: The goat is apparently in pain and therefore needs its help. But the guardian dog sees the herding dog as an extension of the boss and therefore does not attempt to discipline the herding dog. The best situation would be to have both a guardian dog and herding dog who are well managed and experienced at their jobs.

This Anatolian-Maremma cross guardian dog has recently spent some time in the irrigation ditch and is happy to be called to supper.

Here is another livestock guardian dog that only *appears* to be sleeping on the job.

Will a guardian dog fight with trespassing predators?

A guardian dog will bark at, feint at (lurch toward as if to attack), and generally behave in a threatening fashion toward any trespassing predator. Self-preservation is the first objective of all animals, and because actual contact with the trespasser risks injury to the dog, it will generally be used as a *very* last resort.

How many goats can a guardian dog protect?

The type of terrain that the goats are covering is probably the single largest factor in determining how many goats a guard dog can protect. In general, open pastures with good visibility are much easier to guard than rough, brushy terrain.

How many predators there are, and how hungry they are, will also affect how many goats the dog can guard. If the predators, due to numbers or hunger, are more determined to get to your goats and thus predator pressure is high, multiple dogs are almost certainly called for.

Belling (putting bells on some of your goats) can be of great assistance to the guardian dog in doing its job successfully in rough country. If the sounds of the bells suddenly change, the guardian

This Maremma is curious, but as there is no threat to the peace of his charges, he is also undisturbed. *Kaylene Larson, KM Larson Ranch*

dog will be alerted and will rush to the area where it hears the alarm. If you decide to bell your goats, make sure to use bells that can be carried easily and safely by the goats and can be heard from far away.

How much territory can a guardian dog cover?

A researcher of guardian dog behavior and effectiveness in South Africa recently put a GPS unit and activity collar on an eighteen-month-old guardian dog. In twelve hours during one night, the dog covered six kilometers (almost four miles) within one approximately ninety-eight-acre pasture. The dog moved back and forth, frequently crisscrossing the same ground. Why do many guardian dogs often appear to sleep during the day? Because predators are most active at night, and therefore the dog is too.

The collar's tracking device also showed that, following a flurry of activity in one area (the dog was probably barking and running back and forth), the dog briefly left the pasture and promptly (within five minutes) came back. No dead predators were found where the dog had gone out and back in.

The same researcher also conducted a test to determine what the guardian dog would do if an

__Belling__ (putting bells on some of your goats) can be of great assistance to the guardian dog in doing its job successfully in rough country.

The front Karakachan-Maremma pup is confident, but the other would rather look at the photographer from behind its sibling. *Priscilla Ireys, Critton Creek Farm*

UP AND AWAY

In addition to being capricious (being able to change directions very suddenly and without warning at a run) in past centuries, goats learned to climb to evade predators. And although most of us have guardian animals to protect our goats from predators, goats still are very good at climbing, and many seem to do it just for fun. They may not have to escape predators, but they enjoy getting on top of rocks, their mothers, or tree branches.

UP AND AWAY

UP AND AWAY

Carrying food bowls around is a habit of guardian dogs, which they start practicing at a young age. *Pat Fuhr, Giant Stride Farm*

unusual, oddly behaving human tried to get near the livestock. With a camera in hand, the researcher pushed the dog and its sheep around the hill where the sheep were grazing for about an hour. The sheep were closely bunched due to the researcher's presence and the barking and disruptive behavior of the guardian dog. When the researcher decided to run forward toward them, forcing the flock to split into two groups, the guardian dog stayed between the person and the nearest group of sheep, and continued to show aggressive body postures and make lots of noise. The researcher never once was able to touch a single sheep, as the guardian dog prevented him from doing that without risking harm to the researcher or the dog itself.

Goat kids await their mother's return under the watchful gaze of their guardian llamas. *Carolina Noya, Flying D Cattle Co.*

This Anatolian shepherd dog isn't sad, it's listening intently for any sign of disturbance in its goatherd. *Childers Show Goats*

A pair of guardian pups travel in dog crates to their new home on a meat goat ranch. *Ryan Hayes, Timberline Farms*

How long do guardian dogs live, and what can I expect as a normal working lifespan?

How hard a guardian dog has to work, and the care (food, attention, shots) that it receives, will affect its productive lifespan. Typically guardian dogs are long-lived, working well up to or past ten years of age.

If I breed my livestock guardian dog to a herding dog, will the resulting puppies be good at herding as well as guarding?

The puppies resulting from such a mating will probably not do either job very well. The guardian dog that grew up with meat goats will watch, listen, and use its powerful sense of smell to be alerted to unusual things that shouldn't be near the goats, staying near its herd at all times. The herding dog's instinct and training is to make the herd move away from it and to a destination of the handler's choice.

Is there any other way to keep my meat goats safe from predators?

There are alternative methods (some more successful than others) to keep predators from reaching your goats. Night penning your goats, hanging up lights that are said to make predators think they're being watched, fladry (strips of fabric tied to fences that flutter and are supposed to keep predators away), leaving a radio playing during the night, and having guinea fowl near the goats have all kept some producers' herds safe for differing periods of time. One could also hire shifts of armed guards, set up motion-detector lighting systems, and build a very tall fence around your property's perimeter (although eagles don't really care how tall the fence is.) In the long run, however, having such super-max protection pay for itself might be a little more difficult than paying for a guardian dog and its dog food.

In the Resources section, there is information about both Dr. Raymond Coppinger's book and *DogLog*, reproductions of a newsletter about how guardian dogs mature and work.

CHAPTER 8

PREGNANCY AND KIDDING

And now we get to one of the best parts of the year with meat goats! Each season spent helping these wonderful animals thrive is rewarding, but I think that deciding which does will be bred by which buck and then seeing the results during kidding season is like opening presents at Christmas.

You may choose to have your goats kid in sync with nature and have the bucks working at getting the does bred so that kids will be born in spring or summer, in lush green pastures, and bathed in warm sunshine. (And if you think that sounded biased in favor of warm-weather kidding, you're absolutely right!) You may choose, so that you can achieve your goat enterprise goals, to have your goat kids born at a certain time so that the kids grow to a target size for a specific date (a goat show, a 4-H event, or a market-goat customer holiday feast, for example).

If there's an important reason to have your goats kidding during less-than optimal weather, and you have shelter for the goats, it's the right thing to do. But the loss of a single kid due to cold can be a big financial drain unless all of the remaining kids are sold at a significant premium. What time of year you want your goats to have kids is one of many decisions that you should make for a good reason, not just because somebody else tells you to do it.

The white blaze on the face of this cute Boer kid is just a little wider than some others'. That's one thing that makes the kid special, as well as easy to recognize!
Dawn Van Keulen, VK Ventures

PREGNANCY: CONCEPTION THROUGH BIRTH

Breeding Your Goats

There are two options for impregnating your does: natural service (a buck gets the doe pregnant) or artificial insemination (A.I.), whereby previously harvested and stored goat semen is used to impregnate the doe. At the time of this writing, A.I. is possible with goats, but not yet widely used.

A.I. (artificial insemination) can be a good tool in cases where you're trying to ensure reproduction of rare or extremely valuable goats

A.I. can be a good tool in cases where you're trying to ensure reproduction of rare or extremely valuable goats, but if you're simply trying to produce lots of kids, using natural service may be more cost and energy efficient. As you might expect, male goats are not only very good at their job, but highly motivated to do it.

Meat goat does typically have a 21–28-day **heat cycle** and carry their kids for approximately 150 days before giving birth. It appears to most producers that meat goats are fairly seasonal

As breeding season approaches, even does can feel competitive and try to convince each other that they are the boss. *Childers Show Goats*

GOAT COURTSHIP

1. The buck will nuzzle the doe. *Vicky Brachfeld, Half Moon Farm*

2. He will sniff the area under her tail. *Vicky Brachfeld, Half Moon Farm*

3. If the doe is not in standing heat, she will not be interested in his romantic overtures.

4. Or the doe may stand still. *Childers Show Goats*

5. Or the doe may rub against him in return. (Buck smell is attractive to receptive female goats). *Childers Show Goats*

6. She may pee in front of him so that he can smell her urine. He will curl his upper lip up to get a very good sniff to determine if she is in heat.

7. Finally, the buck will mount the doe and breed her.

How do you know when a doe is in heat? She will wag her tail and might butt heads with other does.

breeders. There are exceptions, but most female meat goats have the highest incidence of **estrus** in October through February, and the lowest incidence from April through June. Having estrus, or being in heat, is when the doe produces eggs that can be fertilized by a male goat.

Can you take this information to mean that a doe can't conceive kids in April? No. What it means is that she is *least likely* to be in heat and get pregnant in April, May, and June. After the summer solstice, in late June, does that are not in peak lactation (producing lots of milk for their current kids) will begin coming into heat.

How do you know when a doe is in heat? She will wag her tail and might butt heads with other does. If there is a buck on the other side of a fence, she will be very attracted to him and might rub on the fence and generally behave in a very "attractive" manner to the buck (there are all sorts of ways to describe her behavior—"attractive" is the most polite). When bucks are in rut, they will urinate on their forequarters, and sometimes into their own mouths, as they perfume themselves in order to attract the does. The female goats think that buck smell is very attractive, and they can respond to the scent even if it is brought to them on the wind or on the legs or knees of your chore clothes.

As mentioned in Chapter 4, the urge to reproduce can be very hard on the fences that separate the sexes. An alley or buffer zone between pastures or pens has helped many producers reduce wear and tear on their fences.

So when would kids be born, based on a successful breeding date? Since the time between breeding and kidding is about five months, it's easy to estimate due dates.

If the doe is bred in . . .	Kid(s) will be born in . . .
July	December
August	January
September	February
October	March
November	April
December	May
January	June
February	July

Many meat goat producers who wish to have the highest number of kids born per doe will wait until a doe is in heat for the *second* time, regardless of the month, before putting the buck together with her. Apparently, many female goats produce more eggs in their second heat cycle than in their first. The way that it was explained to me is that the doe's reproductive system shifts from low into high gear the second time she comes into heat. Many producers have complained of unusually low numbers of kids born per doe when they try to have their does bred out of season, or too far away from the October through January period of frequent estrus. Whether the low number of kids born per mother goat is because the doe might have been bred on her first heat cycle isn't known.

One way to improve kid production is to take advantage of the **buck effect**. Buck effect is what happens when a mature buck in rut joins the doe herd in the months that does are coming into heat. Within seven to ten days, you will see many of the does in heat. In some herds, 65–80 percent of the does are successfully bred within the first twelve days of the twenty-one-day heat cycle. Many meat goat producers use the buck effect to have the greatest number of kids born within a concentrated time period.

The buck with these does was wearing a marking harness with a green crayon, and thus the young does that were in heat and were bred have been marked with green on their haunches.

Apparently, many female goats produce more eggs in their second heat cycle than in their first.

If you want most of your does to come into heat for the first time without getting pregnant, you can use what's known as a **teaser buck**. This is a vasectomized buck that's been surgically altered to prevent semen from being ejaculated into the doe. The buck will still act, smell, and behave like a breeding male, but cannot get the does pregnant.

Some meat goat producers have found out the hard way that a vasectomized buck can still have and deposit viable sperm for some weeks following his surgery, so if you have a buck vasectomized for use as a teaser buck, wait a few weeks before you assume that he's safe to use.

When the doe is in heat and the buck is with her, the actual mating can take a matter of moments, or there can be a courtship of hours. The doe usually rubs against the buck, and she may squat and pee in front of him so that he can smell from her urine that she's in heat. She may

The Flehmen response is the way that a buck goat gets the best possible evaluation of a smell, by curling his lip up so that the scent is as close as possible to his olfactory nerves.

play hard to get if she's coming into heat but won't be quite ready for breeding for a few hours yet. Other does, in a very no-nonsense way, will simply back into the buck so that he can do his job. The buck will nuzzle the doe, concentrating particularly on her hindquarters. Quickly, or after much sniffing, snorting, pawing, and turning up of his upper lip in the **Flehmen response** (whereby he gets the scent of the doe as close as possible to his olfactory nerves), the buck will raise his forequarters, resting them on the doe's hips. His penis will extend forward from its sheath. He will insert it into the doe's vulva, pump his hips once or multiple times, and finish with a thrust. You may well see the doe tuck her hips under and forward in a motion that looks as if she had been smacked on her bottom. Don't worry, she hasn't been injured. As my early mentors explained to me, that tucking motion is an instinctual response designed to give the buck's semen the best chance of moving forward into the doe's body and on its way to her fertile eggs.

Either or both goats, the female and male, may well now lose interest in each other, or they may continue romantic behavior for minutes and sometimes hours. Yes, there is quite a range of behavior for meat goats that will still end up giving the same result. Occasionally, multiple female goats are in **standing heat** (ready to be bred *right now*) at the same time. When this happens, the buck has to dispense with romance and get on with it. Does also will compete, sometimes very vigorously, for the buck's attention. This is why many producers separate doe kids under a year old from mature does during breeding season. In our decades of experience, there have never been negative consequences when mature does push and shove each other. However, if a doe kid of smaller size is in heat, and tries to get near a buck, a bigger doe can easily outcompete the doe kid and could accidentally injure the younger female. Choosing to give doe kids the opportunity to get pregnant—letting them have their own buck, or their own time with the buck—is a simple and effective solution to the competition problem.

Since such a large percent of does will be triggered into heat by the presence of an aromatic buck in the first week and a half, some managers of larger goat herds will put multiple bucks in with the does at the same time so that all the does that need it can receive a buck's attention. After ten to twelve days, they remove all of the bucks and put a smaller number of fresh bucks in with the does to ensure that every doe is bred in one 21-day cycle. Be careful if you have enough does in one group to need to use multiple bucks: If all of the bucks are left with the doe herd for the entire heat cycle, as the number of does in heat per day declines, a buck that is breeding a doe can be injured by an idle, and therefore jealous, buck.

This Boer buck has a blue crayon in his marking harness, and his beard is reddish from buck stain. The buck stain on the backs of his front legs is covered by the blue that has rubbed off from does that he has successfully bred. *Bryce Gromley for Gromley Farms*

STUCK IN A RUT IS GOOD, ACCORDING TO FEMALE GOATS

Many buck rutting behaviors might seem odd or less than appealing to most people. While humans compete with each other by showing how smart or attractive they can be, goat bucks crack heads with each other to prove that they are the best buck. Young bucks will sometimes knock heads, trying to establish dominance over each other, until there are drops of blood on the skin at the bases of their horns. Does will also knock heads, but it never seems as imposing as when the bucks do it with all of their weight and determination.

Gestation

As mentioned earlier, **gestation** for a goat kid is approximately 150 days. The does' actual due date (the range is given as 145–155 days) is chosen by

While your doe is pregnant, one of the best things you can do is encourage her to exercise daily.

her body and is dependent on the myriad factors governing when the kids are ready to be born.

While your doe is pregnant, one of the best things you can do is encourage her to exercise daily. Most producers are finding that if the doe gets exercise each day during pregnancy, it will make her labor easier. Exercise can include just walking back and forth to the pasture, or to eat hay outside rather than from a feeder inside the barn. An easy birth will give the kid a better start in life and make the process more normal for the doe.

When your does are heavily pregnant and have bedded down for the night, their breathing may sound just like a chorus of croaking frogs. Don't take this as a sign of distress. It's merely how the does' breathing sounds when there are baby goats taking up lots of space inside of them.

As the birth of her kids approaches, not only should the doe's udder become very full, but her tail ligaments also will relax in preparation for the birth. How can you tell if a doe's tail ligaments are relaxed? On either side of the doe's spine, in the last six or eight inches in front of her tail, the ligaments will temporarily soften, which makes her spine look like it's sticking out from her rump. If you can touch her, her spine will be easier to grasp and you'll feel the individual bones of it more than usual.

Many producers say they can tell when kids have dropped in preparation for birth because the doe's belly looks like an upside-down mushroom rather than a barrel. The bellies of some older does, however, who have already carried many kids in their lives, may well look like upside-down mushrooms throughout their entire pregnancies.

Labor

As labor begins, the doe may well want to walk, and walk, and walk. This may be her body's way of moving the kids into the right position for the birth. Because people care about the doe's birthing of her kids, we want her to be safe and preferably where we can see her. For this reason, many producers will put a doe that's about to give birth in a stall or pen. Some producers like their goats to kid in a pasture or wooded area, attended only by the guardian animal. Many laboring does will leave the herd on their own, some by a few yards and others by a much larger distance.

The doe may paw repeatedly at the ground as if she is preparing a spot for her kids. She will lie down, get up, lie down again, and get up again, sometimes many times. She will probably be calling, but not in an alarmed way (many producers think this is to teach her kids what her voice sounds like). One producer has said that he can tell when a doe is going into labor because she's talking to her belly.

Some does will look as if they're nursing their own udders right before labor. What they're really doing is pulling the **wax plugs** off their teats in preparation for nursing. During pregnancy, the wax plugs keep milk from leaking out and prevent any foreign bodies or germs from getting into the doe's udder. If the mother goat doesn't remove the wax plugs, the kids will remove them when they get their first meal.

Over the years I've learned to accept that despite my desire to help, does really do know best when it comes to birthing kids. It's always tempting to intercede during the birth because the newborn kids are valuable (in time, money, or emotions). But before you rush in, give the mother goat a chance to do it by herself. As long as there is progress happening, assume that she doesn't need your assistance.

How long should a birth take? I really do wish I could tell you a specific number of minutes that

WHEN TO EXPOSE YOUR DOES

In *The Meat Goat Handbook*, you'll find that there's rarely only one right answer to a question. The same thing goes for deciding when to give your **doelings** (young females) the chance to conceive their first kid. This is a decision that should be based on the goals you have for your meat goat operation, since it's *your* business.

Early mentors of mine here in the United States had been taught by goat producers in Australia (with huge herds of goats), and both believed that nature knows best. They gave a seven-month-old doeling the chance to be bred, but typically would **expose** the doelings to a buck who was approximately nineteen months of age and of a bloodline that didn't yield gargantuan kids at birth (cattle producers call such sires heifer bulls, because they've proven that if a cow's first pregnancy and delivery is without complications, the cow has an easier time getting pregnant again). The mentors told me that if the doeling's body had grown well enough in her first seven months of life, her reproductive system would be ready for pregnancy and birth. But most importantly, if she had grown well, her reproductive system would begin having estrus and she would come into heat, without which pregnancy will simply not occur.

I've heard people say that a doeling must be a certain age or grow to a specific body weight before she can conceive. I have also read that estrus is triggered when a doeling reaches a certain percent of the body weight that she will have at maturity. When is maturity, you ask? Ah, another question with two or more right answers. It's somewhere between two and four years of age, depending on the goat's breed, available forage, and environment. Why not trust the doeling and let her instincts and her body decide when she is ready to have kids?

This twelve-month-old Spanish doe is enjoying a Montana summer day with her half-Savanna daughter, born a few weeks earlier.

each stage of delivery takes. In our own operation, we've found that from when our goats have a bag (amniotic sac) showing, within ten minutes the kid is out, and within fifteen more minutes it is up and nursing. Many producers use a 30-30-30 rule (thirty minutes for the labor preparation, the actual birth, and the kid getting up to nurse for the first time). These producers feel that if a doe has been in one stage of delivery for more than 30 minutes, there may be a problem that needs their attention. One veteran producer I know says that if one of her does has been in labor for forty-five minutes and the first kid isn't born yet, she'll assist the doe. Using this rule, she has had to assist only three times in fifteen years and through several hundreds of goat births.

When a doe finally lies down for the last time before her first kid is born, you'll see a glistening, nearly translucent amniotic sac starting to protrude from the doe's vulva. Frequently, you'll see the front hooves of the kid through the sac. The most normal birth position for goat kids, because it's the easiest for the doe, is "the diver." Both of the kid's forelegs are extended, its head is facing forward with its chin resting on its legs, and its rear legs are trailing behind its body. The doe may grunt, groan, and may even arch her body as the contractions of her uterus work on pushing the kid out into the world. Because of the pushing, the kid should slide out of its mother onto the grass or the clean bedding in the birthing area, causing the amniotic sac to

Be they Boer, Spanish, or composite, pregnant does become very wide as their kids grow inside of them. Frequently, does at this stage of pregnancy will make frogging noises when they breathe, which is not a sign of distress.

When the kid passes its first meal of milk as ***meconium*** *(a tarry and black substance), its mother's scent will be reinforced.*

break. At this point, the doe usually gets to her feet, turns, and begins licking the face of her kid (although occasionally she manages to reach the kid's face while still lying down). The licking clears birth fluid from the kid's nostrils so that it can start breathing. If the sac doesn't break during the birth, the doe will rip it by biting at it. The kid should cough repeatedly and will call as it starts breathing.

Now the doe will lick the kid all over as she dries off her new baby. Her licking also stimulates circulation in the kid, and it seems to function like an encouragement to the kid to get going. Within moments, the kid will try to stand up.

If you're watching the birth, this would be a good time to sit on your hands. As I can personally attest, it's hard to *only* watch as the new kid repeatedly tries and fails to stand, then finally manages to stand up, and usually promptly falls over. But eventually the kid should get all four legs under control and manage to stay upright, generally with an attitude of "ta da!" And usually, the energetic licking by the mother goat will knock the kid over at least one more time.

No matter how many times I get to watch a birth, the drive in the kid to get up and find the teat for its first meal never fails to amaze me. Just a few minutes earlier, the kid was floating in a warm, dark world with every need taken care of through its umbilical cord. Now the kid is outside, hopefully in warm sunshine on green grass or in a comfortable shelter, breathing air for the first time, and it knows that it has to find its first meal. This is another time when it's difficult not to help, as the kid may attempt to nurse its mother's knee, her chest, and, if the doe and kid are near one, the wall next to them. But never fear; most mother goats know how to gently direct their kids to where they should be nursing.

Once the kid is suckling, the mother goat will sniff its tail. Sniffing the kid's tail is how the doe knows that it's her baby that is nursing. When the kid passes its first meal of milk as **meconium** (a tarry and black substance), its mother's scent will be reinforced.

There may be a second kid, a third, and sometimes even more kids born in the same short time. This will be another opportunity for the mother goat to impress you with her task management skills and you to practice just watching. But what if there are triplets, and if one of the amniotic sacs doesn't break—should you help? In my personal opinion, you should, absolutely. Tear the sac open and wipe the kid's nose clear (keeping one eye on the mother goat to make sure that she's not misunderstanding your actions). I read once that old sheepherders would use a blade of grass or a piece of straw and tickle the newborn's nostril to make it sneeze and clear the nasal passages. Usually, the arm of my chore sweatshirt does the trick nicely. If you're one of those always-prepared people who I envy, you'll probably grab your handy towel to wipe the kid's face. Once the kid is breathing, take the difficult step backward and away from the doe and kids.

In some instances kids are born backward, or **breech**. If a kid comes out back end first, the doe's labor contractions can push fluid up into its nose. Many kids born breech will cough and sneeze enough to clear their own respiratory system, but if they don't, there's another old sheepherder trick you can use: Grasp the newborn by the back feet and spin it around so that centrifugal force whirls the mucus out of the lungs, windpipe, and nose. We have had to do that, and it works very well. But be forewarned: Keep a *very* good grasp of

HAPPY AND HEALTHY GOAT BABIES BOUNCE!

Sometimes as early as a few hours of age, kid goats will suddenly pop up off the ground for no apparent reason. Don't worry, there's no stray electrical charge under their feet—the kids are just healthy, happy, and bouncy!

Vicky Brachfeld, Half Moon Farm

the kid's back feet, as the kid will be soaked with incredibly slippery birth fluid. A washcloth would be a wonderful, lightweight item to carry in your kidding kit, and it would offer you a good grip on a wet kid if the need arose.

"The diver" might be the way kids are typically born, but you'll see kids come out in other birth positions sooner or later. Does can *usually* deliver a kid even if it's in an abnormal position, but the birth might be much harder for her, and thus also for the kid. If you're there, and you see that progress isn't being made, now is the time to help.

If you have to get your hand inside a goat, it's best to clip your fingernails short and remove any rings with protruding settings. You can carry a small bottle of lubricant in your kidding kit—it's rarely needed, but it is always much appreciated by both the goat and the helper. (Wearing an examination glove will help prevent possible disease transference between you and the goat.) If you are alone and have to help a doe, slipping a collar or halter with a lead rope on her and tying her will help make sure she stays in one spot. Gently slip your clean and well-lubricated hand inside the doe's vulva, keeping your fingers together and palm turned in toward the center of the animal. The various birth positions that we know of are illustrated on page 153—try to ascertain which one you're feeling inside the doe. You're trying to get hold of both of the front legs, and if the head is forward, you will gently but firmly pull the kid out by the front legs. Whether the doe is lying down or standing, pull the kid slightly downward toward her **hocks** (back knees) as you slide the kid out. As the doe has contractions, she will clamp down on your wrist/forearm. Breathe deeply (or try humming) and don't let go of the kid. The doe will relax in a moment, and you can finish gently and smoothly sliding the kid out. Good job!

If we have to help a doe by getting a hand inside of her, afterward we'll give her a shot of penicillin and some time alone with her kids in a place where the whole herd doesn't come trooping through. We call it a TLC (for tender, loving care) pen. It's nice for the new mother, but I think that

THE BIRTH OF A GOAT

Illustrations by Trevor Burks

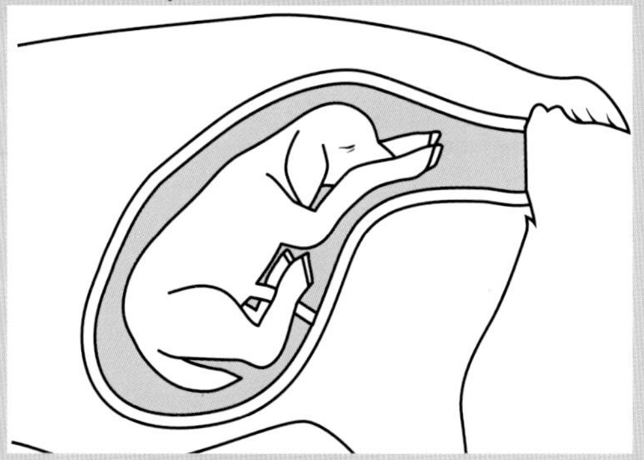

1. Normal "diver" birth position

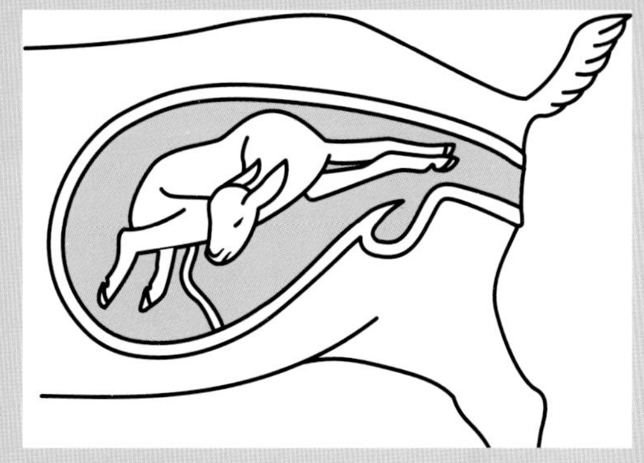

2. Head back

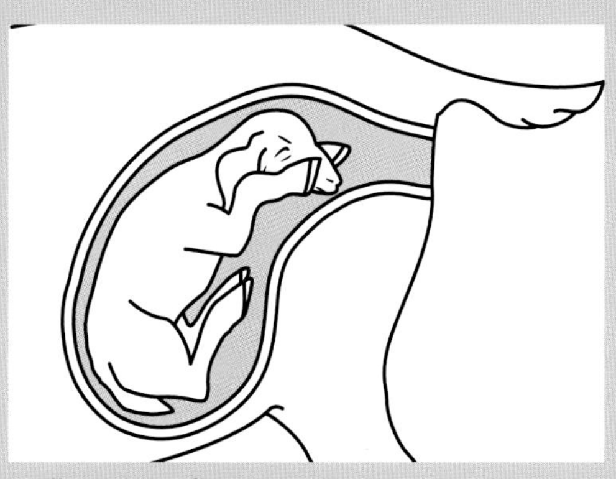

3. Elbows bent, feet next to nose

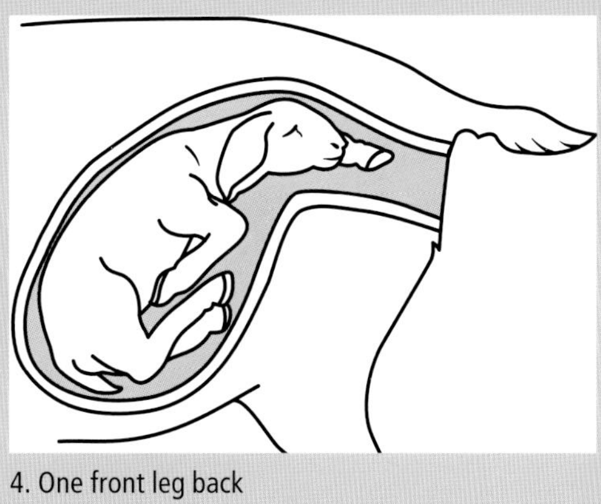

4. One front leg back

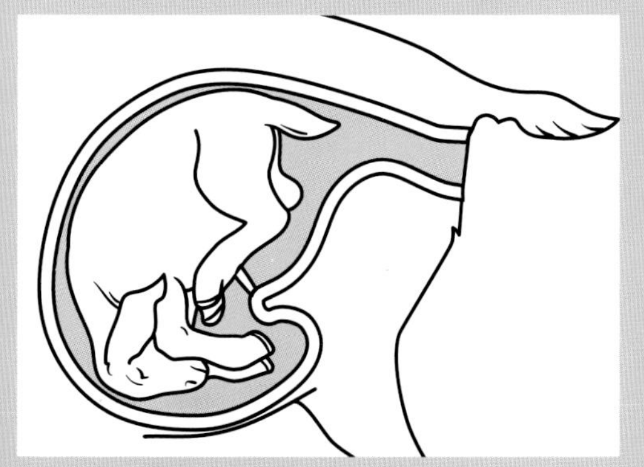

5. True breech

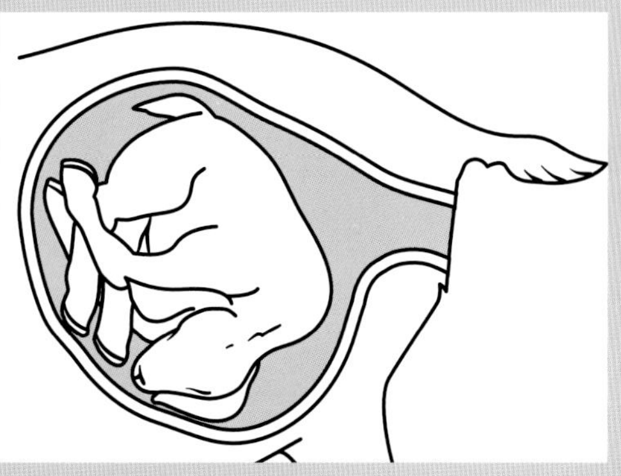

6. Cross-wise

THE BIRTH OF A GOAT

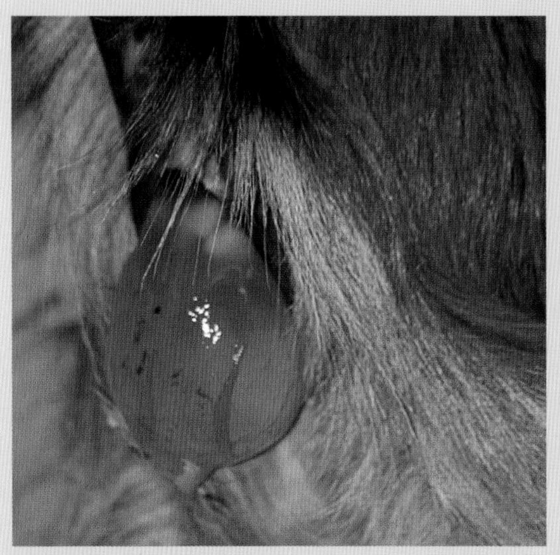

The glistening amniotic sac that surrounds the kid and contains the fluid that has protected the kid during its development is the first thing you will see during delivery. *Vicky Brachfeld, Half Moon Farm*

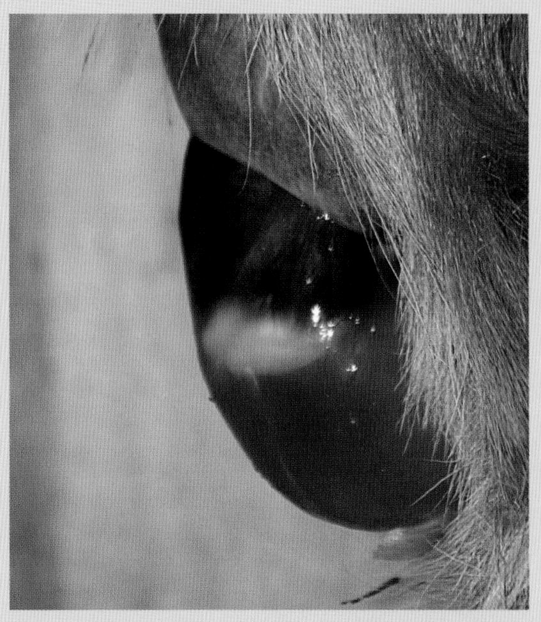

Through the sac, the front feet of the kid are visible. *Vicky Brachfeld, Half Moon Farm*

This doe is lying down, which means the kid has to fall only a few inches before landing on soft bedding. Some does, while busy licking their first kid, are standing when they deliver their second kid, who ends up with a much farther drop! *Vicky Brachfeld, Half Moon Farm*

The birth sac is tearing open naturally, and now the kid will start breathing air for the first time. *Vicky Brachfeld, Half Moon Farm*

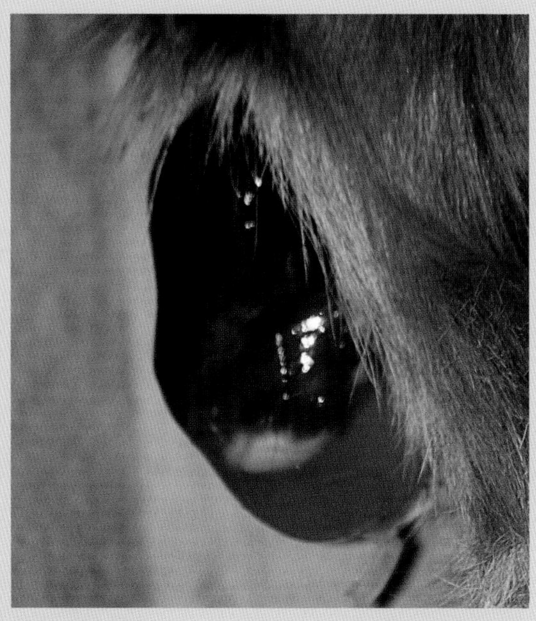

Then the head comes, resting on the forelegs. This is an example of the typical "diver" position of delivery. *Vicky Brachfeld, Half Moon Farm*

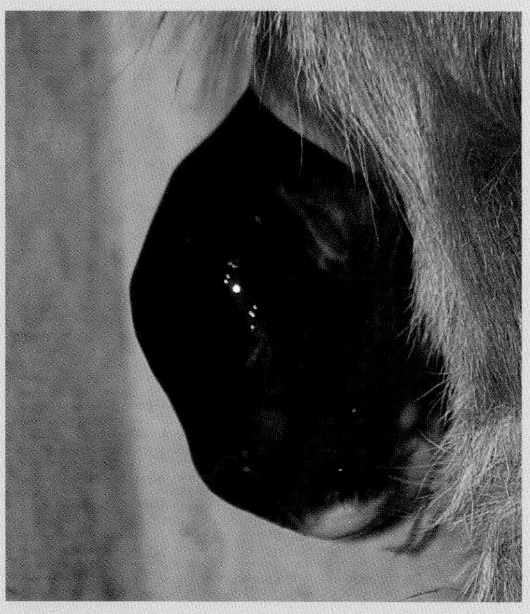

Once the head is outside of the mother goat, the rest of the kid's body typically follows quite quickly. *Vicky Brachfeld, Half Moon Farm*

The doe usually gets to her feet to begin the task of licking the kid. The licking both dries the amniotic fluid from the kid (wet kids can become chilled, which is not good) and stimulates the kid's circulation and activity (which is good!). *Vicky Brachfeld, Half Moon Farm*

The mother enthusiastically licks the kid's ears, face, and whole body. *Vicky Brachfeld, Half Moon Farm*

The afterbirth, while unappealing to humans, is full of nutrients that new goat mothers need.

putting the doe and kids where we can check on her easily is mostly for *our* benefit.

Some meat goat producers dip the navel of newborn kids in iodine to ensure that no bacteria on the end of the umbilical cord can travel into the kid's body. If the kids are born outside on clean pasture rather than in a poopy barnyard, or on clean new bedding in a kidding stall, iodine is less necessary.

After a doe has delivered her kids and they've nursed for the first time, the doe will pass or deliver her afterbirth. The afterbirth is a pinkish soft mass with darker patches or spots in it. The spots are where there was an attachment to the inside of the goat's uterus.

And, oh my! What is that goat *doing*? If she's like many goats, she's probably eating her afterbirth. The afterbirth, while unappealing to humans, is full of nutrients that new goat mothers need. Not all does eat the afterbirth, however—some let guardian dogs clean it up for them. In our own operation, we make sure that the guardian dog doesn't move in to eat the afterbirth unless the mother goat shows no interest in it.

After watching the big goats eat hay, a young kid tries some to sees how it tastes. *Vicky Brachfeld, Half Moon Farm*

Milk from Mom is the best. *Dawn Van Keulen, VK Ventures*

AFTER THE BIRTH

Evaluating Maternal Instincts

All does should welcome their new offspring. If one doesn't, check her physical condition. If she is thin, her body knows that she will have a hard time raising one or more kids because there are inadequate body reserves (fat and muscle) to draw from for nursing. It seems to many producers that there is a direct link between the doe's udder and her love for her kids. If her udder is full because she received adequate nutrition while her kids were growing in utero, the doe will be entranced by her kids. If her udder is slack due to hunger, a high parasite load, or an illness, she will obey nature's first rule, which is to ensure her own survival before successfully raising the next generation of her species.

From my perspective, the best mother goats will **milk down** when raising her kids. This means that her body will use up fat stores when she is making milk for her kids. As the kids grow and start to forage

First-time mothers, particularly young females who are still growing themselves, usually have just one kid.

Most meat goat does make wonderful mothers, as this Spanish doe in Pennsylvania demonstrates. *Marcus Briggs, Dancing Heart Farm*

for themselves, their demand for milk from the dam (their mother) declines. If pastures are adequate to the needs of the animals, the mothers will easily regain their weight in time to come into heat and be bred for the following year's crop of kids.

If the mother goat's physical condition is normal, yet she chooses not to be a good mother, the problem might be mental. Has another goat or the guardian dog interfered in the birthing to the point that the real mother has been driven away from her kid? (Remember that licking the newborn kid is the job of the mother goat, not the guardian dog or another doe.) Have humans selected attributes in the goat's family tree other than maternal qualities to the point that the mothering instincts have been eclipsed or lost? Any of these things might cause attachment problems.

As you get to know your goats and their strengths and weaknesses, you can decide what type of mother goat you would prefer to have and whether maternal attributes are more important to you than the kid's growth rate, appearance, or pedigree. Some first-time kidders, only a year old themselves, try to bite us when we ear-tag their kid, and to us, those are fantastic, protective goats with exceptionally good maternal behavior.

In a perfect world, every doe would have excellent maternal attributes, and each kid would always be healthy and grow well. This might not be a perfect world yet, but that's still an excellent and achievable goal for your goats. Your objectives, environment, and the type of lives that your goats lead will determine what your perfect goat's behavior and performance is.

Does love their kids. Taking a nap together is well earned after a successful delivery. *Childers Show Goats*

This doe has given birth to twins, a common result of a caprine pregnancy. *Kim Hunter, Fossil Ridge Farm*

This mother has delivered triplets in different colors. *Weed Goats 2000*

Litter Size

First-time mothers, particularly young females who are still growing themselves, usually have just one kid. In their second and subsequent pregnancies, does typically produce twins, triplets, and, less frequently, quads. Very rarely, a doe will have quints or sextuplets.

The mother goat usually has two functioning teats on her udder, but she can comfortably raise triplets and even quads if she can eat enough food for her body to make milk for all her kids. As a good manager, you will keep an eye on the litter of kids to make sure that they're all getting enough to eat from their mother and that the smaller of the kids aren't being outcompeted by their siblings.

If it appears that one of the kids is noticeably falling behind the others in rate of growth or

Four kids from one mother is less common, but possible, as evidenced by these quadruplets. *John and Chad, Willow Valley Farms*

vigor, you can either take it away from the mother to raise yourself as a bottle baby, or take a bottle to the kid once or twice per day to supplement its mother's milk. If the kid really needs the extra food, it will learn very quickly that you have a snack and will come running to you for the milk. This way of helping a very fertile doe will also help the kid grow up while learning how to be a goat.

Typically, each of the kids in a triplet or quad birth will not be as large as single or twin kids of the same age, but if you add up the weights of all of her kids combined, that doe will prove to be one of your best producers.

Born a quad herself, this Boer-Spanish doe produced quads. *Kaylene Larson, KM Larson Ranch*

AH-CHOO!

Goats make a warning sound that very much resembles a sneeze. So if you are walking through your goatherd one fine summer day, and something tickles your nose and causes you to sneeze, do not be surprised if many of the goats jump to attention, or jump out of your way! Even a bottle baby kid, who is about as tame as a goat can get, jumped and looked at me with great attention when I sneezed in her presence.

This young doe in Wisconsin is still trying to lick her new kid dry, and the kid is ready to start exploring. *Kim Hunter, Fossil Ridge Farm*

Cold Kids Are Slow Kids

If a newborn kid appears listless and isn't fighting to get up and nurse, it might be too cold. Here's an easy way to determine if the kid's core temperature is low: Slip your finger into the center of the kid's mouth. (I find that my pinkie finger is most sensitive to temperature.) If the inside of the kid's mouth is cold to the touch, get the kid into a warm water bath before it fades away.

Keep the kid's head above the water level as it soaks—the utility sink in a laundry room is a wonderful place for warming kids. Recheck the kid's mouth temperature until it feels warm to the touch (the kid may well start moving and trying to suck at this point). Before you take the kid back to its mother for its first meal, make sure it's completely dry so it doesn't chill down again.

Some people say to put the kid in a plastic bag from the neck down to keep it from getting wet, the theory being that water will wash the doe's scent off the kid, causing her to reject it. I've never witnessed a doe not taking back a warmed-up kid, but if you're worried about it, there is one thing you can do—if the doe passed the afterbirth while you were warming the kid, you can rub the afterbirth on the kid's back, and especially the area around the kid's tail. The doe will start licking the kid, the kid will start nursing, and all will be as it should be.

First Milk

The first milk that kids get from their mother is called **colostrum**. Typically, it's a thick, slightly yellowish milk. It contains all of the antibodies that a newborn kid needs, along with a variety of things that trigger different fuctions in the newborn goat's body.

Many meat goat producers call colostrum the elixir of life, as it is so critical for newborn kids to ingest if they're to survive and thrive. The

THE FIRST MEAL

Welcome to the world, baby! This Boer doe in Oregon knows exactly what to do for her newborn. Licking the kid and "talking" to it, she encourages her kid in its quest for the first meal of colostrum. *Childers Show Goats*

The kid isn't quite ready to stand up, but crawling works too. It is definitely headed in the right direction. *Childers Show Goats*

The kid searches for breakfast. (Who else is wanting to help, just looking at these photos?) *Childers Show Goats*

Yes! Mealtime and, wow, is it delicious. The kid is not going to let go any time in the near future. *Childers Show Goats*

Well done, Mom! You're the proud mother of a beautiful set of twins. *Childers Show Goats*

EXPERT TIP:

Cut bottle nipples open vertically, *not* horizontally, and only make a tiny opening. Remember that when the kid works at sucking the milk from the nipple, the saliva that the kid's mouth makes starts digesting the milk. Place the nipple in the kid's mouth so that the valve is clear and can allow air back into the bottle. When you wash the nipple after each use, make sure that the small metal ball bearing that makes the flutter valve work doesn't accidentally get out of the notch where it should be.

doe produces colostrum for the first few days of lactation. The newborn kid's digestive system can absorb colostrum only for the first thirty-six hours of its life, after which time the walls of the digestive system change so that the colostrum can no longer pass through them.

Kids are sometimes unable to drink colostrum due to weakness or abandonment by their mother. Many colostrum replacement products are available for meat goat producers, but none are as perfect as the real thing. Harvesting some extra colostrum from does that have plenty is an inexpensive way to have extra insurance on hand (milking some colostrum out is also fairly easy, especially with a helper). Freeze batches of colostrum in small zippered bags to keep it handy. When you need it, defrost the bag in a warm water bath rather than a microwave—a microwave will neutralize many of the antibodies in the colostrum.

There are a number of ways to get colostrum into kids who need it. We find that the vast majority of kids who need extra food will easily take milk from a nipple. The best nipple we've found is called a Pritchard teat. The base of that nipple, which can be screwed onto most empty twenty-ounce plastic soda bottles, has a flutter valve that allows air back into the bottle while the kid nurses so that the vacuum the kid creates as it sucks the milk doesn't flatten the bottle or make it very difficult to draw the milk out. With warm milk in the bottle (test it on the inside of your wrist), place the nipple in the kid's mouth and push a few drops of milk through the nipple by squeezing gently on the bottle. Usually getting a taste of milk in the kid's mouth is all it takes for the kid to start vigorously sucking on the nipple.

Some producers will tube feed a kid by sliding a thin rubber hose down its throat, putting milk in a syringe, and using the syringe and tube to gently push the milk into the kid's stomach. Sounds easy, doesn't it? I'm sure that for some people it is. We've never done it because we've never had a kid fail to start nursing from a nipple. We're also fearful of pushing milk into a kid's lungs accidentally, which would be fatal.

A teacher of mine has explained patiently (and repeatedly) that you can tell if the tube is in the kid's esophagus by feeling with your fingertips: If you've accidentally put it in the windpipe, the

Almost all meat goat does are supportive of their kids. Both of this doe's kids play on her back in Idaho. *Weed Goats 2000*

bands of cartilage that keep the windpipe round and open will also prevent you from being able to feel the tube. You can work with the kid's swallowing reflex to ensure that the tube is in the right place, and double check by sucking through the tube once it's in the kid. You should suck up stomach fluid; if you only suck air, you've landed in the lungs and will need to rethread the tube.

HEALTHY KIDS

You can help your does raise their kids by making it possible for the does to feed their kids well and teach the kids how to be goats. If goats have the three S's available (sustenance, shelter, safety), they will use what they need and do their jobs very well. You can choose to do many things to or with the kids, but what will happen if you don't do something? We find that only the failure to neuter male kids (or wean them from their mothers) will have lasting effects that may not be what you wanted to have happen. Do goats (especially kids) need toys? The goats say yes and the owners who enjoy laughing at the kids' antics agree.

CHAPTER 9

RAISING KIDS

Once the kids have been born, the goat's job in raising them is to provide them with nutrition in the form of milk and to teach them how to be a goat. Goat knowledge imparted by a mother includes how to eat solid foods (and which foods to avoid), how to run for shelter in case of bad weather, how to stand up on one's back legs to eat leaves, and what to do when the guardian dog barks a warning (get in a close group with the other goats, stand right behind the dog, and look where the dog is looking).

Your role in raising kids is to make it possible for the doe to feed and teach her kids. Whether her feed grows in a pasture or comes in a bale, there should be adequate quantity and quality to let her body make the milk the kids need and for her body to start replacing the resources that it used during pregnancy. The doe shouldn't have to worry about predators. (That statement is, in fact, a matter of opinion. Some meat goat producers accept losses of kids to predators as a fact of life due to the location or extent of their pastures, but I believe that producers should do what we can to make life safe for our animals.)

Your role in raising kids is to make it possible for the doe to feed and teach her kids.

Doing what you can is exactly that—it's doing what you *can* do. For example, you won't be able to keep tornadoes or earthquakes away from your goat pasture, but it *is* achievable for you to give your goats access to adequate feed, water, and minerals, and to have a guardian animal to keep the goats safe.

So what are some of the things you might do for the kids once they're born? Following are a few of the most common possibilities.

If you let a kid in the porch, don't be surprised when it jumps up on a chair! *Amber Leininger, Striking A Livestock*

This doe in Texas is just trying to keep an eye on her own kid, but the mothers of all the other kids are happy to let her babysit their kids while they eat. *Carol DeLobbe, Bon Joli Farm*

This brand-new kid has to take a nap, right here and right now, so its mother stands guard. *Carol DeLobbe, Bon Joli Farm*

BOTTLE FEEDING

As we learned in Chapter 8, kids can be bottle fed if they're unable to drink from their mother's udder (for whatever reason!). The main argument in favor of bottle feeding a kid is that, without it, the kid may die or, out of desperation, attempt to attach itself to another doe. Since goats don't readily accept kids that aren't theirs, the hungry kid may well cause the chosen doe to run away from it and her own kids as well.

The arguments against bottle feeding are that the kid isn't worth the time and energy it will take to raise it and the cost of the milk replacer or fresh milk is a cash drain. I'm not taking sides on this argument, other than to say that bottle babies do demand time, but when they grow up, they are usually exceptional goat ambassadors.

When the kids come running to nurse, the doe turns her back knees out and usually burps up a cud. *Freddie Brinson, Brinson Pineywoods Cattle and Spanish Goats*

In beautiful Maine summer weather, this doe and her kids enjoy an afternoon nap together in their shelter. Choosing to sleep in a shelter has to do with the goats' love for safe places, even those partially enclosed. *Marge Kilkelly Dragonfly Cove Farm*

A happy medium if you want the goat to grow up happy but don't have the time to feed it yourself could be to find a young or retired person who wants to spend time feeding kids for the fun of it. The young person might be one who wants to start a meat goat operation and has time to invest in helping you in return for a goat or two. A retiree may well have the time and inclination to spend on bottle-feeding kids, but may not want to keep the goats long-term.

Bottle baby goats, like all young animals, need to be touched by a caregiver in order to grow up happy. Normally, the mother goat gives her kids that physical contact naturally. If you choose to bottle raise a kid, the kid will want and need more from you than just milk so many times per

THE REWARDS OF HAVING YOUR GOATS RAISE KIDS

What's the best thing about getting to watch your does raise their kids? The joy you feel at their birth? The excitement of seeing them start their lives? The anticipation of being able to show or sell them, or watch them reproduce? To me, it's all of the above. Raising young kids is yet another wonderful part of the year with meat goats. I love watching the kids explore their world, then suddenly zoom back to Mom as though they were checking up on *her*. It's all part of being around these entertaining animals. Sometimes a young (or young at heart) doe will even join in the kids' cavorting.

After nourishing her kids, a doe shows them how to be goats. Just who is showing whom here? *John and Chad, Willow Valley Farms*

Meat goats are very affectionate creatures. *Dawn Van Keulen, VK Ventures*

day. But watching a goat kid start bouncing with joy because you're coming and, even after it's fed, continuing to love you, is priceless.

Where can you find the milk to feed or supplement a kid? There are commercially available (at the feed or farm store) bags of kid milk replacer to make goat's milk out of. There may be a goat owner nearby who milks his or her goats and has milk for sale that you could purchase and use to feed the kid. Some producers even buy milk from cattle dairies and use it to feed bottle babies. However, cow milk is not quite as rich as goat's milk, and sheep milk is more fatty than goat's milk, so goat milk replacer is the best for a kid.

Using fresh (or frozen) goat's milk and cow's milk could possibly bring diseases into your goat operation that you don't already have, whereas a bag of goat milk replacer could be believed to be clean of diseases. When you mix the contents of a twenty-five-pound bag of goat milk replacer according to the directions, it will make just over seventeen gallons (2,176 fluid ounces) of goat milk. So the cost of a bag of kid milk replacer divided by seventeen is how much making a gallon of milk from powder will cost you.

So how much milk should you feed a kid? Start with a bit less than—and over a few days, work up to—10 percent of the kid's body weight in fluid ounces of milk per day, divided into at least three servings. So, a _____ -pound kid x 0.1 = _____ pounds of milk per day. Multiply that number by 16 (ounces per pound) = _____ how many ounces of milk the kid needs per day. Divide by 4 to get how many ounces to give the kid per serving, if you can feed the kid a breakfast, lunch, dinner, and late evening feeding. For example, you could spread the kid's meals out and feed it at 7 a.m., noon, 5 p.m., and 10 p.m. After a week or two, most kids can get a third of their milk in three servings per day.

And *please*, do not give in to the kid's beautiful, pleading eyes and insistent nosing and nuzzling and let it eat its fill. You can, in fact, kill a kid with kindness by overfeeding it *much* more easily than you can starve it. The serving size can increase as

Meat goat producers would call this scene a kid pile. The kids sleep while the does eat so their bodies can produce milk for their babies. *Kaylene Larson, KM Larson Ranch*

Is your ear edible? *Kim Hunter, Fossil Ridge Farm*

An upside-down plastic pot with a rounded opening cut into one side makes for a perfect place for a meat goat kid to sleep. *Jesse Bennett, Driftless Land Stewardship LLC*

the kid gains weight for the first month, then hold it steady even though the kid is still growing rapidly.

A very kid-friendly schedule would be

Day 1: divide feed into 6 servings
Day 2–14: 4 servings per day
Week 3–4: 3 servings per day
Month 2: 2 servings per day
Month 3: 1 serving per day
Month 4: no more milk

Even young goat kids do sleep at night, and with a late-evening meal they will be quite content until the morning. Here's a warning to both ladies and gentlemen: bottle babies do know where the milk is *supposed* to come from and will bunt (push very suddenly, and very hard) at your chest if you bend over to pet the kid or pick it up. A kid's

This Boer doe's teats are kid-friendly, just the right size and positioned perfectly for a hungry kid to nurse. *Carol DeLobbe, Bon Joli Farm*

instinct is also to bunt for milk above its head in the dark, which can be unexpected and painful to your groin if you're crouching or sitting.

Bunting is good, according to a mother goat, as it's a signal to her to let down her milk so that the kids can nurse. You will see, as the kids come running toward her, a mother goat turn out or open the knees of her back legs to give her kids good access to her udder. As the kids bunt at her udder and start nursing, she will sniff their bottoms just to make sure that they're her kids, then burp up a cud and start chewing contentedly.

For how many weeks do bottle babies need to be fed? There is a range of answers to this question, but most producers see their does tapering off the frequency and duration of their kids' nursing to nothing by the time the kids are three months old. An early mentor of mine had an answer that was easy to remember: Feed three times a day for the first month, two times a day for the second month, and once a day for the final, third month.

If you're going to use concentrates (grains) as part of your goat diet, start offering them to the kids after you see them chewing their cud.

If the doe's udder has just one teat on each side, and no false or extra teats, when the kids latch on to nurse, they will get a good flow of milk. False or extra teats typically have a limited flow of milk. *Dawn Van Keulen, VK Ventures*

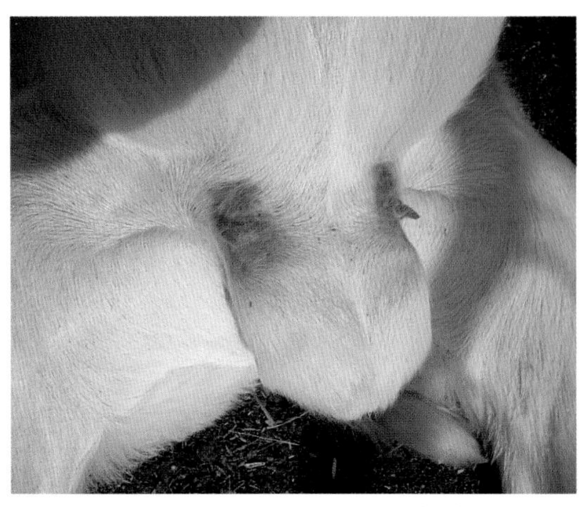

Even though he will never need the teats for feeding kids, the teat structure on a buck is very important. Since every female goat that a buck sires will carry his genetics, many meat goat producers select bucks with just one teat on each side of the scrotum. *Dawn Van Keulen, VK Ventures*

We stop giving the bottle babies milk when they're three months of age. The kids should start nibbling at grasses (probably dry ones) within a week or two after birth. By four weeks of age, many kids are chewing their cud. If you're going to use concentrates (grains) as part of your goat diet, start offering them to the kids after you see them chewing their cud. Most producers who bottle feed kids or include grains in their goats' diet do carry out a vaccination program including shots for clostridium perfringens types C & D, more simply called **enterotoxemia** or overeating disease. When a goat gets too much of a food containing carbohydrates, be it milk or concentrates (grain), bacteria normally present in the digestive system in small quantities suddenly proliferate. The toxins they produce when in such abnormally large numbers poison the goat and will cause it to die painfully. Prompt application of antitoxin can save the goat. It's a whole lot kinder to the goat not to get the last few sips of milk or just one more handful of grain than it would be for the animal to suffer with enterotoxemia.

Make any changes in a goat's diet slowly, so the digestive tract can function well by having the time to produce the necessary digestive enzymes.

TAGGING

Tagging goats as a means of identification is something that many meat goat producers take as a given (and that sentence did come out sounding rather decisive, didn't it?). A small plastic tag, usually approximately one inch square, is placed in the kid's ear at birth to accurately identify which kids belongs to which mother. Tags can also be applied later when kids are ready to be sorted into "keepers" and "market kids." The accuracy of linking kids to mothers by that time would be much lower than at birth, but kid production per doe might not be something those producers are tracking.

Arguments for tagging include the fact that ear tags are fairly inexpensive, quick and easy to apply, and usually stay in the goat's ear for life. Many goats change in appearance as they age, but the numbers on an ear tag are printed into the plastic and may fade, but will not change.

This curious kid has been tagged with its own ID number. The innate curiosity of goats makes them want to find out what things are, and this characteristic also can make it easier to handle them. *Leslie Edmundson, Little River*

Arguments against ear tagging include that some producers don't need a number on their animals to recognize the goat accurately. Others find that the cost of the tags and the time needed to apply them aren't worth it if they don't need to know the details about which does are raising which kids. Those producers can make good decisions based on the total number of kids available for sale and the total weight of those kids.

If a producer wants to keep records on individual goats, the ear tag number is a very easy code to use. Lists of ear tag numbers can be sorted, organized, and reorganized quickly and easily. Ear tags can carry your farm logo, or the first digit in the tag number can identify the year that the goat was born. Many producers tag all of their female goat kids in the right ear, males in the left, as a handy visual aid for sorting animals by sex as they come down the handling alley. You may say that you will always recognize every goat by sight and that may well be true, but for those of us who can't do that, ear tags are a very handy tool!

Ear tags can also carry a **scrapie number** on them, which is required to get a health certificate from a veterinarian. A health certificate is required for goats to be transported across state lines legally for breeding, or if the goats are going to an auction or to a processor. **Scrapie**, a fatal degenerative disease, is almost always found only in sheep, but goat producers have asked to get and use a scrapie number on their ear tags to assist in eradication of the disease.

Many producers tag all of their female goat kids in the right ear, males in the left, as a handy visual aid for sorting animals by sex as they come down the handling alley.

To tag a goat, the ear tag is placed in the applicator with the male half of the tag's pointy tab facing the round opening in the female half. A dab of antibiotic ointment can be applied to the tag's point, then the two sections are pressed together through the goat's ear. The goat will feel a pinch as the point penetrates the ear, but once the two tag sections click together and you let the kid go, very few even shake their heads or seem otherwise concerned with their new jewelry. The mother goat frequently takes more issue with the decoration of her beautiful babies, sniffing and licking the tags, giving you what really appear to be dirty looks.

TATTOOING

Tattooing is required if you want to register your goat with one of the different breed registries. A tattoo gun (really more of a pincher) is used to make small holes in the skin where ink is then rubbed. This puts an indelible unique code inside the goat's ear or in the skin under the tail.

The main arguments for tattooing are that it's permanent and can be seen by the human eye. Arguments against tattooing can include the pain caused to the goat in applying the tattoo or that the tattoo could be altered after application by a dishonest person. But if you wish to register your animals, and a tattoo is required, you will have to tattoo your animals.

As a side note, microchips also can be used for permanent identification. The chip is embedded in the goat's skin and is read by a hand-held machine that is waved over the chip.

NEUTERING

Some decades ago at a meeting of fledgling American cashmere goat producers, an Australian researcher got the attendees' attention by saying, "The lust of the goat is the bounty of God." The phrase was not only very memorable, but we have all been reminded of it when our bucks prove it by trying to "do their job" with such conviction and concentration. Before I got my first bucks, I was told that bucks only come into rut in the winter. Once I had bucks, I realized that winter, according to bucks, can be anywhere from early July until late June the following year.

Buck kids can be viable by ninety days of age. Not all are, but many will be. Viable means that when a buckling practices breeding behavior and the other goat is a female in heat, she can and probably will get pregnant. And bucklings will start practicing at a very young age. Many meat goat producers have seen bucklings flipping their forelegs and making cricket noises at their sisters within hours of birth. It's entertaining when they're that young (all right, *some* meat goat producers have a creative sense of humor), but as the months slip past, it's easy to keep thinking, "Oh, he's still young, he's not *really* doing anything, he's just practicing." Well, practice makes perfect, and separating the bucklings "next weekend" is sometimes too late. Mark your calendar for ninety days after the *first* kids are born and get intact buck kids away from the does, or neuter male kids well before ninety days to ensure there won't be any surprises five months later.

What are the arguments for neutering? Neutering is the only perfect method of preventing a male goat from impregnating a female goat, other than putting a truly goat-proof fence or distance (at least half a mile, according to our experience) between the sexes. There are plastic or leather aprons that people in other countries put right behind the forelegs of their male goats

Along with practicing their climbing and bouncing, meat goat kids start practicing reproductive behaviors at a very young age. Luckily, both kids are a few months short of reaching puberty. *Kim Hunter, Fossil Ridge Farm*

to physically prevent bucks from being able to impregnate does, but I haven't seen them in the goat catalogs yet. I would worry about the straps getting chewed through by a romantic female goat in order to remove the buck's apron.

What are the arguments against neutering? Well, growth rates in intact male kids are greater than growth rates in **wethers**. That's not the whole story, however: When a buckling starts rutting, he puts the energy that was making him grow so well into other things, such as showing other goats that *he* is the best buck for the job, and his growth rate slows. Many producers are leaving their male market kids intact until approximately three months of age, then neutering them; others never neuter the male kids, selling them intact.

Does the market reward or penalize sellers for offering intact males? Some consuming groups actually prefer bucks, many do not, and some are unbiased, so there's no single answer to that question either. An older (close to or more than one year of age) intact male goat will rarely grade as a Selection 1 animal, because if he has a surplus of nutrition, he will happily use his extra energy for showing off his muscles, not adding to them.

The methods for neutering male kid goats include **banding**, "Burdizzo-ing," and cutting. Any of the methods will render the kid infertile by making it impossible for the sperm to travel up through the cords connecting the testicles to the body and get ejaculated during breeding.

In banding, an elastrator (a tool like a reverse-action plier) is used to stretch a small, circular rubber band open. Once the band is placed around the narrowest part of the skin between the goat's body and its testicles, the elastrator is removed. The rubber band disrupts the blood flow to the testicles, and within two weeks the small, dried-up scrotum will drop off.

Here's why we're biased in favor of banding: We do it within a day of birth for the vast majority of our buck kids. Very, very rarely has a kid reacted negatively to the banding, most running back to their mothers for a meal as soon as the band is on and the kid is released. Banding can also be fairly easily done by one person, alone.

The **Burdizzo** is a tool that severs, by clamping with strong pressure, each of the two individual cords in the "neck" of the scrotum, between the goat's testicles and the body. Each cord must be clamped for approximately twenty-five seconds. I am told that using a Burdizzo is a two-person task. There is an excellent presentation of how a Burdizzo is used online, and the way to access it is listed in the Resources section.

A male goat also can be neutered by physically cutting open the bottom of the scrotum and removing the testicles. If this is the way that you choose to neuter a kid, please realize that the process is an invasive one—you're making an opening in the skin of the goat, and skin is the protective layer that keeps things as small as bacteria, and as large as flies, out of the inside of a goat.

I realize that no matter how I attempt to rewrite those sentences, they come out sounding rather biased against the procedure of cutting. I will say, however, that a friend surprised me by saying that she cuts her buck kids by herself, sprays an iodine solution on the opening, and releases the kid with no negative results.

When veterinarians perform surgical neutering, they anesthetize the animal, ensure that surfaces and tools are sterile, and then neuter the goat. The goat is therefore not at risk for pain, infection, or blood loss due to incorrect technique. Your vet would probably be happy to perform surgical neutering for you. Your vet or mentor may also be happy to teach you how to use an elastrator or Burdizzo, or as I learned recently, cutting.

Arguments in favor of cutting include safety and a pain-free process if a vet performs the operation and speed and a low cost if you perform it yourself. Arguments against cutting include the potential for pain on the animal's part, the difficulty of performing the operation, (that is, the emotional difficulty of cutting into your animal's body without the benefit of anesthesia), and the risk of infection.

WEANING

Weaning is the physical separation of the kids from the mother goat, which, because of the separation between them, makes it impossible for the kids to nurse from their mother. For bucklings that are going to stay intact, weaning by ninety days of age is a necessity. I don't mean to sound heartless, but most ninety-day-old bucklings don't miss their mothers when they're weaned. Their mothers don't seem to miss the bucklings, either.

Whether you decide to separate female kids or neutered male kids from their mothers is a choice that only *you* can make! Many producers wean their kids as a matter of course, saying that mother goat needs to have the kids removed so that the nursing stops, and thus the drain on her system stops as well. Some producers don't wean, and their does regain plenty of body weight to breed another crop of kids one year after the last crop. So, yet again, there are different ways to achieve both the goal of having the goat kids grow well and the goal of the mother goat regaining enough weight to start the next pregnancy.

Here's an opinionated statement, backed by twenty years worth of experience: When you don't wean, the herd stress level drops dramatically, as do health problems caused by stress and containment problems (challenges to fences). The noise level also drops from pandemonium to nothing. When kids and does are forcibly separated from each other, they don't understand the reason, and they'll frequently call to each until they're hoarse.

For bucklings that are going to stay intact, weaning by ninety days of age is a necessity.

Is there a way to wean meat goats that's low stress? I believe so, and it's done by the mother goat. When we see a kid just over ninety days old with its mother, and the kid starts to try to nurse, the doe simply moves her back leg *into* the way of the kid, instead of out of the way as she did during the first months of the kid's life.

DISBUDDING

Disbudding, or dehorning, is the process of applying a hot iron to a goat kid's head to keep its horns from growing. We never have our meat goats disbudded for a number of reasons. Horns are a normal part of the goat's body, are believed to be part of the goat's cooling system, are part of how goats vanquish thistles, and are what the goats use to scratch their own back. There's also the fact that it's much more comfortable for the goat and handler if the handler can grasp the goat by the horns in order to restrain it if necessary (rather than, for example, grasping the goat around the neck, which can trigger panic).

Who usually disbuds their goats and why? Some 4-H and other shows require goats to be disbudded in order to be shown. Many meat goat producers simply decide that it would be better for the goats (or for them) if the goats didn't have horns. Owner or goat safety is a factor, as well as convenience in feeding.

If you choose to disbud your goat kids, it should be done within the first one or two weeks of the kid's life. Please call a friend who routinely and successfully disbuds his or her kids or a dairy goat producer who does the same. The calmest and most matter-of-fact online resource for information about disbudding is listed in the Resources section.

CREEP FEEDING

Creep feeding is a method of offering kid goats extra feed that the mature goats don't have access to. Typically the way of enclosing the area where the creep feed is available will allow the smaller kid goats to creep through small openings or

Goat kids grow well under the care and training of their mothers. This young Kiko doe's kid is growing just fine.
Ernest Wohlford, Wohlford's Farm

between slats in the fence, but prevents larger goats from getting through.

Arguments for creep feeding include giving the kids extra rations in order to maximize their growth, possibly achieving a certain size by a certain time for a sale or show. Arguments against creep feeding include possible lack of cost effectiveness (the kids may grow more quickly, but the higher sale price received doesn't cover the cost of additional feed) or having an enterprise goal of optimizing animal growth on pasture rather than by feeding grain.

How would you make a creep feeder if you chose to offer the smaller goats extra feed? The Texas A&M publication cited in the Resources section recommends five inches of space between the vertical bars of the creep gate and an adjustable vertical bar to prevent taller but slim mature animals from entering the enclosure.

WEIGHING

Weighing kids can be a way to monitor the birth weight and growth rate of the young animals. This would give you information you could use to make decisions for your goat enterprise, if growth rate is one of the factors that you wish to track. If you want or need to know how much your kid weighs, using a scale is the easiest and

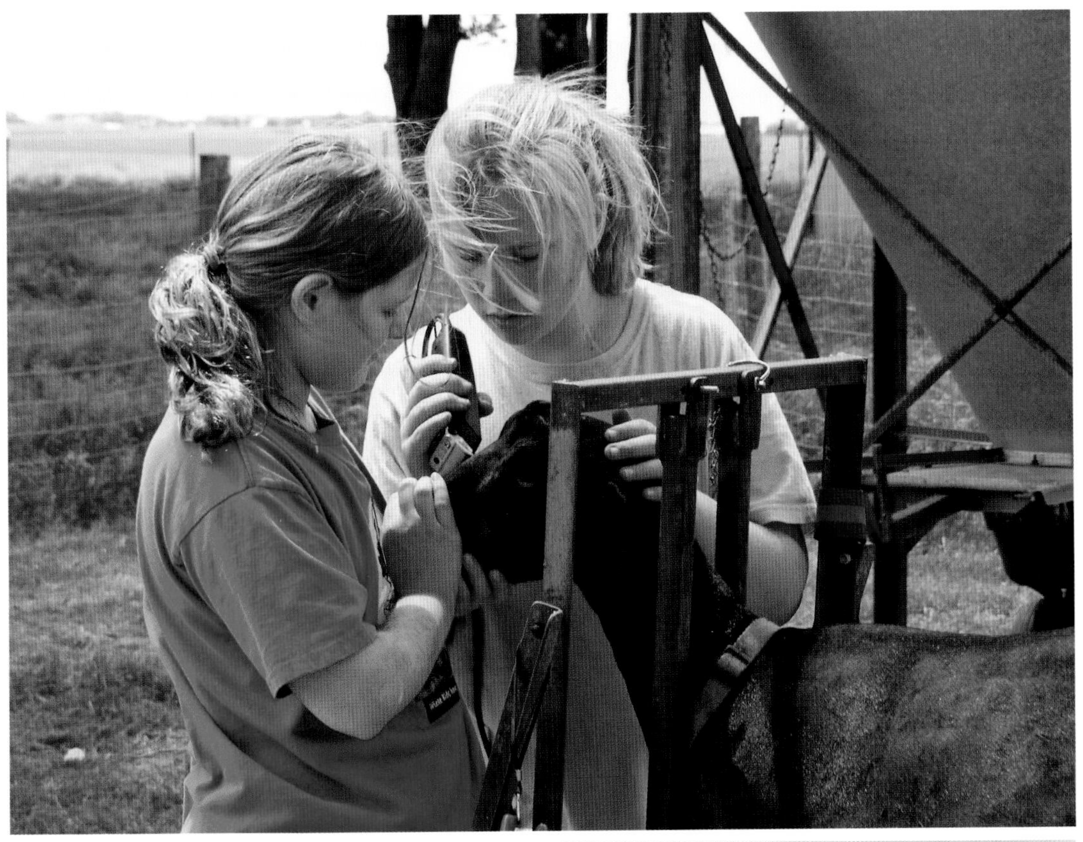

Preparing goats for show day includes trimming and grooming each animal, also called fitting. *Osceola County 4-H, Crazy Ideas Farm*

most accurate method. A hand-held fish scale and a nylon or canvas sling will work quite nicely for young or small kids. As kids get bigger, you can take the bathroom scale outside, with a board to use as the floor under the scale. Weigh yourself, and then pick the kid up and weigh both of you together. Subtracting the smaller weight from larger one will give you the weight of the kid. A helper who can see the numbers on the scale would be of great assistance.

Eventually the kid will get too big for you to hold or for the scale to weigh both of you together. If you still want to find out how much the goat

Teaching the goat to stand well for the judge's inspection is one of the responsibilities of the handler. In showing a goat, the owner tries to present it to the judge in the best way possible. *Richard and Sandy's Boer Goat Farm*

weighs, you can either ask the vet if you can stop by and use his or her scale (your goat had better lead well so that you get it from the front door to the examination room where the scale is). Or, you can start shopping for scales!

If you've already thought about or constructed your handling facilities, hopefully you've included a place to put in a scale in the section of the alley where you can temporarily hold one goat. Many producers have very successfully repurposed hog scales to weigh their goats; this can be very cost-effective. These scales can sometimes be found as surplus equipment at a farm or farm store. There are also new scales available, in both digital and dial versions.

Why is it good to be able to weigh your goats? Here's one example: Let's say that two mother goats live on the same farm and have access to the same feed. One of the does raises two 80-pound kids, while the other doe raises two 70-pound kids. In this situation, the first doe would usually be considered superior to the second one whether you sold the kids at market or as breeding stock, right? Ah, but what if you were able to weigh

Many producers have very successfully repurposed hog scales to weigh their goats.

the two does, and found that Doe 1 weighed 145 pounds and Doe 2 weighed 105 pounds? If you wanted to know the reproductive efficiency of each doe, you could now calculate it:

2 x 80 = 160 / 145 = 1.10.
Thus Doe 1 has a reproductive efficiency rate of 110%.
2 x 70 = 140 / 105 = 1.15.
Thus Doe 2 has a reproductive efficiency rate of 115%.

So which one is the better doe? If a major goal of your goat enterprise is reproductive efficiency, Doe 2 is better. But what if Doe 1 has other traits besides reproductive efficiency that make her more valuable to you? Or what if you're considering the goats' financial efficiency? In the example above, Doe 1 would require 38 percent more feed per day (and thus per year) than Doe 2. So Doe 1 will give you 14 percent more pounds of kids per year, but will require 38 percent more feed to do it.

In suburban Arizona, meat goats carry out prescriptive grazing. It's easier and quieter to have goats "mow" the narrow yards. The goats are fun to watch at their work, and they fertilize while they remove unwanted vegetation. *Brad Payne, Arizona EcoGoats*

VACCINATIONS

For goats, no medicine in the world can compare to a healthy start in life. When a mother goat has adequate nutrition during pregnancy, and isn't unduly stressed by weather or predators, her kids have the best chance of being born with all the building blocks they need for good growth and long, healthy lives.

Many meat goat producers carry out carefully planned vaccination programs for their goats and kids. To them, the effort and cost of doing vaccinations is well worth it. Conversely, some meat goat producers don't vaccinate their animals at all and yet still have healthy, productive herds. Whether vaccination would be more or less necessary for your herd depends on whether you frequently take your goats to other places, or other goats frequently come to your place. Contact with outside goats increases the exposure to strange (to your goats) bacteria. Since your goats have no immunity built up against these bacteria, having rarely or never encountered them, they're more likely to become infected. Goats that travel to shows or exhibitions should therefore be vaccinated regularly, as their need for protection will be much greater than that of their homebody counterparts.

RECORD KEEPING

This isn't actually something you need to do for your kid goats, but rather something you can choose to do for your goat enterprise. Records of your animals' weights, health, vaccinations, and any other facts that intrigue can be written down in a notebook (unless you have nearly unlimited, accurate memory). Dictaphones are also very useful, portable, and slip easily into a pocket. The goats also are not as liable to taste a cassette recorder as they are the sheets of a notebook (but either record-keeping method can accidentally be dropped in a water trough, with poor results). The information recorded on a dictaphone would need to be changed into a format that you can read, compare, and reference after you're done observing your animals.

TOYS

Some people may wonder why goats need toys. In an effort to be open-minded and diplomatic, I have to say that goats may not, in fact, *need* toys.

Do goats need *toys? Maybe.*
Do goats enjoy *toys?*
Absolutely.

Okay, that's enough open-mindedness; I'll let the goats tell you why! Just watch your kids play on their mother, a ramp, a teeter-totter, a hillside, an empty barrel, the guardian dog, a fallen tree, rocks, surplus cement steps (*that* would be recycling, with the immediate and visible benefit of hoof-trimming done by the goats themselves). Do goats *need* toys? Maybe. Do goats *enjoy* toys? Absolutely.

Someday, we may know for certain that goat kids with toys grow up healthier than those without

Without the bucks, there would be no goat kids! A Boer-cross buck and a pure Spanish buck in Canada are ready and willing to go to work. *Myrna and Walter Coombs, Llone Pine Farm*

LEFT: A Boer kid sleeps peacefully after a hard day's play. *Dawn Van Keulen, VK Ventures*

toys. Until then, I'll make a creative suggestion: Goat owners whose goats have toys are healthier and happier than goat owners whose goats don't. Why? Because laughter has been proven to be good for people, and it's unlikely that very many people can watch a goat kid play on a toy without chuckling or laughing out loud.

CHAPTER 10

SELLING MARKET GOATS

Selling market goats can be another great time of year with meat goats, if the kids that you bring to market sell at a good price. When you get the check from a buyer or sale facility, it can feel like getting a very good grade on a test or a paper at school.

When are your market goats ready to sell? Well, do you simply want the goats off your property?

Goat meat, called *chevon*, is popular with ethnic markets. This is a Portuguese meal of grilled chevon. You may choose to sell cuts of *chevon* direct to the consumer, or you may choose to sell live goats at auction. *Shutterstock*

A public auction yard serves as a meeting place for many buyers and many sellers of market meat goats. Here, a group of market kids arrives at the sale. *Carol DeLobbe, Bon Joli Farm*

Or are customers clamoring for goats? If you *have* to sell, you are not in as strong of a position to negotiate as you would be if you're sitting back in your recliner, feet up, looking through current and historical sale prices to decide when you *want* to sell.

Let's say that maximum prices for market kids for the last few decades have typically peaked around Easter, while minimum prices have typically bottomed out in mid- to late summer. Furthermore, let's assume that you have market kids ready just before Easter, and they sell for $2 per pound. If you had spent $1.50 per pound of kid to raise them to market time, feeding them through the fall and winter, is that sale ($2 - $1.50 = 50 cents per pound net) better than selling kids in mid-summer at $1 per pound but having only 25 cents per pound of costs invested in their production, because they have only been grazing and browsing, and haven't needed to eat any purchased feeds? Actually, the mid-summer sale is more profitable ($1 - 25 cents = 75 cents pound net). Good for you! Profitability analysis might be a very important-sounding term, but it's fairly straightforward and easy to do.

By **backstrapping** each market goat, you will know how many good, fair, and poor quality animals there are in the group. By weighing them, you will know how many pounds of good, fair, and poor goats you have. Now *you* are in control of selling your goats in the easiest way that will yield you the most money possible. You could even call two or three different buyers, and say, "We have so many kids that are Selection 1, and their average weight is so many pounds. We also have so many kids that are Selection 2, and their average weight is so many pounds. What are you interested in paying us per pound for each of those groups of goats?" Remember to ask if there is a difference in price per pound and how much the difference would be if the buyer came to your farm or ranch to pick up the animals versus you taking the goats to the buyer at an agreed-upon location.

Knowing how much the difference in your revenues would be will allow you to figure out whether you or the meat buyer can travel more cost effectively. Keep in mind that if the buyer comes to you and wants to haggle over price, you could just smile and ask if he or she would like another cup of coffee or a doughnut (you wouldn't really be trying to negotiate, just trying to be hospitable and polite). Whereas if you took your goats to the meeting location, you would now have the time and gasoline invested in getting the goats there.

By backstrapping each market goat, you will know how many good, fair, and poor quality animals there are in the group.

Pens of market goats wait for their turn to be offered for sale. *Producers Livestock Auction Company, Inc.*

SELECTION 1, 2, 3

Not too many years ago, when a goat was brought to an auction facility, a processor's harvest plant, or a parking lot where one person had a goat and another person wanted a goat, the buying and selling was a matter of convincing the other person of how much the goat was worth. Now, we have a nationally accepted numerical grading system to identify the quality of an animal offered for sale. Just as the cattle industry uses the terms "prime," "choice," and "select" for A, B, and C or top, middle, and bottom grade, the meat goat industry now has Selection 1, Selection 2, and Selection 3.

Those of us in the meat goat industry can thank Dr. Frank Pinkerton, Dr. Ken McMillin, and Rebecca Sauder of the USDA Market News Reporting System for not only seeing a need for us to have a grading system for meat goats, but creating the terms and descriptions that have now become widely adopted. They are the meat goat industry's unsung heroes. What would happen if we didn't have Selection 1, 2, and 3? Just imagine trying to convince a police officer you weren't speeding if your car didn't have a speedometer and the police didn't have radar.

Processors, auction facilities, and most order buyers are especially happy that there is now a numerical grading system for meat goats.

Processors, auction facilities, and most order buyers are especially happy that there is now a numerical grading system for meat goats. One processor, who has been harvesting meat goats for the grocery stores in a major metropolitan area for decades, said that "before we had the Selection 1-2-3 terms, buying meat goats was a real free-for-all."

Does that mean that your opinion of your goats is absolutely the same as the opinion of the buyer? No. It just means that you can use the same terms to describe the market goat; it doesn't mean that you will be in complete agreement.

So what exactly do Selection 1, 2, and 3 meat goats look like? According to the Louisiana State University Agricultural Center, the categories are as follows:

- Selection 1 goats have "superior meat conformation with thick muscling throughout the body that will give a high meat-to-bone ratio. Moderately thick muscling appears through the chest, and the muscling over the back strip is full and rounded. The outside leg has bulging muscling, and the outside shoulder is moderately thick."
- Selection 2 goats have "average muscle, the chest muscling is moderate, and the back strip muscling is flat, reflecting slight fullness along the back. The outside leg has only slightly thick muscling, and the shoulder muscling is slightly thin, with a medium meat yield, because of the average muscle-to-bone ratio."
- Selection 3 goats have "inferior meat conformation. This goat has slightly thin muscling through the breast and along the back and

Buyers can look at the goats that will shortly be offered for sale before they choose to bid on the animals. These goats and buyers are just outside of the sale ring. *Carol DeLobbe, Bon Joli Farm*

very thin muscling through the leg. The legs, back and shoulder are narrow compared with the body length. The sunken appearance at the top of the shoulder, below the loin, top of the rump and base of the leg indicates the lack of conformation and poor yield of meat from this goat."

(Descriptions cited with permission from Louisiana State University Agricultural Center Publication 2951, 1/08 revision. Please see the Resources section for information on how to download the publication.)

Is there a handy way to tell what selection your market goat is? Yes, there is!

Dr. Sparks's knuckle test (adapted by permission)

If the short ribs (behind the rib cage but in front of the hip bones) feel like . . .

- the back of your hand near the wrist, which is to say that you can tell there are bones inside the goat's body but the bones are not distinct, that goat is probably a Selection 1 (carrying a "good" amount of flesh).
- the knuckles when your hand is stretched out almost flat, that goat is probably a Selection 2 (carrying a "medium" amount of flesh).
- your fingertips when your fingers are stretched out flat, the goat is probably a Selection 3 (carrying a "poor" amount of flesh).

Pretty "handy," right?

The goats are brought into one half of the semicircular ring. Buyers indicate their bids to the auctioneer. *Producers Livestock Auction Co., Inc.*

THYME FOR GOAT

Thyme for Goat, a partnered group of Maine family farms, is proud to offer naturally raised, humanely harvested goat meat available for shipment directly to the customer's door, offering an alternative to the factory farm–produced meat available at most supermarkets. All of the goats are raised in a healthy, natural environment. The goats have free access to pasture, browse, and get plenty of fresh water and sunshine. Grain is only fed as a supplement to ensure a balanced diet. Hay, pasture, and browse make up the majority of the goats' feed.

Thyme for Goat has opted not to go fully organic because the group believes that it's not in the best interest of the animals. It doesn't use hormones of any kind, but it does reserve the right to treat its goats for parasites and to use antibiotics when medically necessary. Thyme for Goat believes that these methods achieve a balance that provides the customer with a high-quality, wholesome product, and provides the goats with healthy and productive lives.

Thyme for Goat's animals are treated with respect from the day they are born, and the end-of-life process is as low stress as possible.

Thyme for Goat offers a variety of frozen *chevon* products, including sausage links, boneless shoulder roasts, cubed stew meat, loin on rack of ribs, pepperoni, salami, and much more. Says a company representative: "There is nothing better than coming in from your favorite winter activity and curling up by the fire with a hot bowl of your favorite comfort food. Goat is an excellent choice of meat for making curries, soups, casseroles, and chili." (www.thymeforgoat.com)

TO MARKET, TO MARKET

There are many ways to sell your market goats, ranging from public auctions to direct to processor sales to direct to consumer sales. Selling doesn't have to be hard; in fact, it might be as simple as making a telephone call and delivering goats when they're needed. Below is a rundown of the different options for selling goats.

As you read this list, keep in mind that the law of supply and demand is our friend.

Historically, sale prices for market goats have been lowest in the summer and highest in the winter, due to abundant supply of market animals being available in the summer and limited supply being available in the winter. Speaking as a meat goat producer, the fact that many customer groups do not prefer to purchase frozen meat is good for us. That way, ocean-going ships with containers that hold thousands of pounds of goat meat are not typically a competitive source for the product that we raise.

Auctions

When you take your goats to an auction facility, you turn them over to professionals who make a living selling livestock. The auction doesn't work for you, and it doesn't work for the buyers. The auction is the location where buyers come to find the animals that they need to purchase. The auction facility presents groups of goats to the buyers and

The auctioneer is seated centrally above the ring. As the representative of the auction facility, he serves as the intermediary between buyers and sellers. The goats flow very quickly from one side of the ring to the other. Buyers have only moments to offer their bids before the goats are no longer available. *Producers Livestock Auction Co., Inc*

makes a commission from each animal sold. The commission isn't a great deal of money, and therefore the auction facility needs to sell as many animals as possible each sale day. Most auction buyers have spent years looking at and bidding on meat goats, and they wouldn't have a job for long if they spent more than the goats were worth. They have enough experience to judge the selection of a group of animals, and depending on what their customer is looking for, will spend only as much as is necessary to purchase the animals. Neither the auction facility or the order buyers are there to keep *you*, the goat producer, in business. They are certainly not there to cheat you, either; they're just there do their jobs as effectively as possible.

Direct to Processor Sales

Some processors will buy animals directly from the producer. This would eliminate one or more of the people between you and the consumer. Selling your goats directly to a processor can take a little more work and make you a little more money, or it can require a lot more work and make you a lot more money! Is the processor there to keep you in business? No, the processor is there to turn meat goats into goat meat and supply their customers (usually grocery stores) with the products they want. So it is in the processor's interest, if you have the goats that their customers need, to keep in contact with you and to make sure you're a happy vendor.

A traditional holiday meal for many cultures is roasted chevon. *Shutterstock*

Curried goat is popular in Caribbean, Indian, and Southeast Asian cuisines. *Shutterstock*

Goat meat can be made into low-fat sausage links. Chevon has fewer calories and less saturated and unsaturated fat and cholesterol than beef, pork, and even chicken with the skin removed. *Shutterstock*

Cultures on the Indian subcontinent use chevon in their cuisines, such as in this mixed tandoori platter of goat and chicken. *Shutterstock*

Direct to Consumer Sales

Some meat goat producers really enjoy working with customers. Other producer cooperatives such as Maine's Thyme for Goat work for the benefit of all members *and* the good of their customers. All of the joys and challenges of people working together, including getting into an agreement and staying in agreement on goals and responsibilities, how to fine-tune or alter the group's actions, and how to get along while doing it, are inherent in any type of group effort. And yes, a partnership with your spouse is a group effort!

Direct to consumer sales often mean selling market goats at a livestock auction or direct to a processor. But other possible avenues for selling your meat goat products include cooperatives, farmer's markets, and CSAs (community supported agriculture). According to LocalHarvest, an organic and local food website, CSAs are "a popular way for consumers to buy local, seasonal food directly from the farmer."

HARVESTING THE MEAT

For almost two decades of raising meat goats, no matter how much I realized that meat goats became goat meat, I managed not to see a harvest happen. And especially because I enjoy the animals, I thought that it would be really hard to watch. But I have to tell you, it was a relief. I got to see how a **halal** facility treats and harvests the goats, and what was so amazing is how instantaneous the actual death is. One moment the goat is standing there, and the next it's simply not alive anymore. A halal harvest is done by slitting the goat's throat, and I have now learned that unconsciousness is instant due to loss of blood pressure. But before I learned the "why," it was very clear to me that the transition from goat to carcass was extremely fast and virtually painless. In the hours of watching the processing plant in operation, there were no animal calls of distress, no panicked animals, and the plant didn't smell bad at all.

From the time when a goat walked from a holding pen to the harvest floor until the carcass was hanging in the cooler took approximately 20 minutes. In that time the goat is killed; the blood is allowed to drain out; and the carcass is trimmed, skinned, eviscerated, washed, inspected by the USDA inspector, weighed, tagged, and moved into the cooler.

GOAT POWER!

So what's so special about goat meat? Only everything! Goat is the healthiest meat you can eat. Just look at the numbers from the USDA, taken from the "What's in the food you eat?" search tool on www.ars.usda.gov (data given is for a 100-gram serving).

Nutrient	Goat Baked	Chicken Baked without Skin	Pork Chop Broil Lean Only Eaten	Beef Prime Rib Lean Only Eaten
Energy calories	142	188	208	198
Protein (g)	26.99	28.69	28.33	28.32
Fat, (g)	3.02	7.35	9.72	8.52
Cholesterol (mg)	75	88	78	75
Saturated Fat (g)	0.926	2.023	3.61	3.228
Unsaturated Fat (g)	1.354	2.638	4.413	3.531

This Goat Kabobs wagon is seen at many venues in and near Virginia, and it does a very good business selling delicious cubes of goat meat cooked on a skewer. *Renard Turner, Vanguard Ranch*

A goat mentor told me that her husband, who does the processing of their goats, feels that the halal method of harvest is definitely the easiest on the animal. Using a firearm (typically a pistol) could be considered easier for the person, as the person is to some degree separated from the animal, but is not as instant and easy for the goat.

STEP-BY-STEP HALAL PROCESSING

- When the goats arrive at the processing plant, they are unloaded into holding pens, which have shelter, food, and water.
- When the kids are needed on the harvest floor, two goats walk together over floors with good traction. Walking with another goat keeps both animals calm and happy. Why good traction? When an animal does not slip, it is not afraid (of falling) and its muscles stay relaxed and do not produce adrenalin.
- The goats stand side by side in the harvest enclosure with the processor standing behind them. The processor holds one goat's head, cupping the goat's chin with his hand, and slits the throat of the goat in a single, smooth motion. Note: a *halal* or acceptable to Muslims processing facility is one from which meat products can be consumed by those of Islamic faith and also by people who follow Judaic **kosher** food preparation guidelines. The person who does the slitting of the animals' throat must be a Muslim and is typically (but is not required to be) a man.
- The goat is instantly unconscious due to loss of blood pressure and drops to the floor. Even before the second goat can react to the movement of the first goat, it has also been dispatched.
- The blood from the goats' carcasses drains from their bodies.
- After the carcasses have drained, they are placed on a kicking table, where very normal post-death movement of the (primarily back) legs can occur.
- The carcasses are trimmed (feet and horns removed), skinned, and eviscerated (inner organs removed).
- The carcass is skinned. The workers in the plant in which the pictures on the following pages were taken use a knife to skin the carcass, as their customers prefer minimal fat to be left on the carcass. The other way to skin a carcass is to use air to inflate the skin and then pull the skin off, but that method leaves more fat on the carcass.
- The carcass is placed on a hook that hangs from the rail, is washed, weighed, and tagged (the weight is written on a tag that is pinned inside the body cavity).
- For the final consumer's protection, the USDA inspector checks over the carcass. If the inspector deems anything unacceptable (blemishes in the muscle from trauma or an incorrect treatment method of the goat having been given injections, or abcesses in the liver from the goat having had liver flukes), the carcass is taken back and the unacceptable muscle or organ trimmed. The carcass is then reweighed, reinspected and, if passed, placed in the cooler.
- When the carcass has cooled to the proper temperature, it can be taken by refrigerated conveyance to the grocery store and sold to the final consumer.

WHY PERFORM A QUICK AND PAINLESS HARVEST OF ANIMALS?

No matter what standard or faith one follows, it is the right thing to do, morally and ethically.

It keeps muscle oxygen levels high, resulting in a more tender meat with a more pleasing color (if the animal is not stressed, no muscle activity happens to form metabolic byproducts and cause tightening in the muscle tissue).

It facilitates a rapid and complete bleed-out of the animal, which is important for meat to have good color and flavor characteristics as well as to be in compliance with cultural or religious mandates.

HALAL MEAT PROCESSING

Two goats walk together on a short ramp with nonslip flooring, freely and without being prodded. Walking with another goat keeps both animals relaxed and content, and having secure footing keeps the goats from tensing in fear of falling.

The person who dispatches the goats stands immediately behind the animals. (This person must be of the Muslim faith and is traditionally a man.) A blessing is said for the spirit of the animal, giving thanks for the nourishment that the animal will provide. The goat's throat is slit quickly and in a single motion from side to side with a razor-sharp knife. The animal is instantly unconscious due to loss of blood pressure, and as the blood drains from the body, all body systems fade to a stop. The transition from goat to goat body takes less than a second.

An international report states that "religious slaughter is the most humane way because it leads to less trauma to animals" (The French Ministry of Food, Agriculture, and Fishing's "Bibliographical Report on Religious Slaughter and the Welfare of Animals," 2008). The rapidity and totality with which the animal's body empties of blood ensures that the meat fulfills Halal as well as Kosher dietary requirements for health and cleanliness, offering consumers optimal color and flavor characteristics at the same time. *(Photography by Mike Plavchak)*

Goats arrive at the processing facility in a variety of conveyances ranging from goat totes and crates in the back of pickup trucks to trailers and semis. The goats are unloaded into shaded and well-ventilated pens (and have protection from cold winds in the fall and winter).

The pens always have water available for the goats, and the animals are fed if they will be held in the pens for more than a few hours.

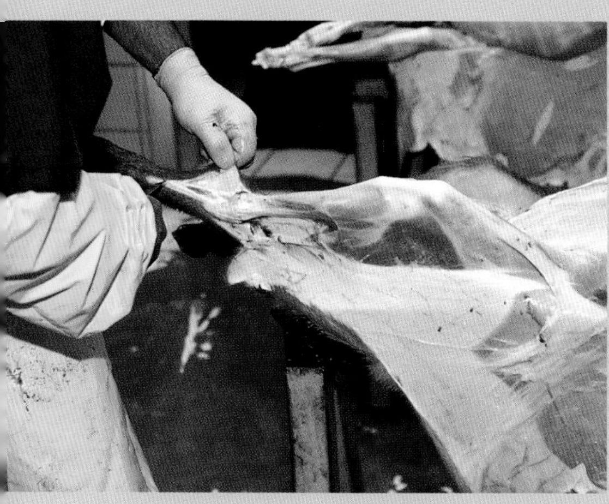

Animals harvested at this plant are skinned by hand, rather than having the hide stripped off by machine. Customers prefer manual skinning because it leaves minimal fat on the carcass.

Internal organs other than the liver, kidneys, and heart are removed from the carcass, and "kidney fat" is scooped out and discarded due to customer desire.

The carcass, now classified as "on the rail," is thoroughly washed before traveling across a scale and being inspected by the USDA.

The inspector determines if the carcass is acceptable to be sold to the consuming public, checking to ensure that there are no abscesses in the liver, for example. (If the inspector finds abscesses, the liver is condemned and removed from the carcass.)

The carcasses are chilled in the cooler until they reach 40 degrees or below, the safe temperature for refrigerated transport to the point of sale.

A refrigerated truck brings the goat meat to grocery stores for customers to buy and enjoy.

COOKING WITH GOAT

We have repeatedly heard from people who try goat meat for the first time that they're pleasantly surprised by the delicious flavor. Unfortunately, your local grocery store doesn't always sell goat meat (yet!). If that's the case, you can always order some *chevon* online and cook it yourself or take a goat (market kid) to a processing plant to be transformed into goat meat. Whether you or someone else will butcher the animal, you should get or ask for the following cuts:

front shoulder—as roasts, or cubed for kabobs

loin—have cut in medallions (small chops that are wonderful barbequed) or as a bone-in rack of kid

hindquarters—roasts

shanks and ribs—as is

trim meat—have ground for burgers, shepherd's pie, meatloaf, or sausage, or cubed for stew meat

When purchasing goats for meat, keep in mind that approximately 25 percent of the animal's live weight will be bone-out meat.

Once you have your cuts of goat meat, why not have a group of goat friends over to cook and sample dishes in a buffet-style or tasting station gathering? Invite your neighbors, your mentor, the vet—anyone you think might enjoy the great taste of goat (don't forget the vegetables, salads, side dishes, and desserts). For each dish, write out a small card with an explanation of what the dish is so that people can know what they're trying. Some nonthreatening dishes include meatloaf, shepherd's pie, kabobs (small cubes of meat threaded onto a skewer alternating with chunks of vegetables), a nice roast with potatoes and gravy, and, for more culinary adventuresome folks, curried dishes ranging from mild to hot. And don't forget goat's milk ice cream for dessert: You can either buy this treat prepackaged, or make your own using goat's milk and an ice cream maker.

Are you getting hungry yet? I know I am! Just describing the kabobs made my mouth water. If you're craving a goat meat meal, try one or more of the recipes below. Your family (and your stomach!) will thank you.

OVEN GOAT SOUP OR STEW

This is the perfect one-dish dinner for busy evenings.

2 pounds goat stew meat
4 medium onions (peeled, but leave root end on)
3 garlic cloves
4 medium potatoes
4 large carrots (cut in two)
1 teaspoon rosemary

In a heavy Dutch oven, preferably cast iron, combine all ingredients. Fill pot with water about halfway. Place pot in oven at 250°F–275°F. Cook for 4–6 hours.

CARRIE'S CROCK POT CHEVON

Two small to medium goat roasts (frozen)
Cream of mushroom soup (condensed)

In the morning, place the frozen goat roasts in a crock pot, mix the soup with water as instructed, and pour over roasts. Turn the crock pot on low and put the lid on. The roasts will be ready for dinner. *Author's note:* This recipe is easy enough for *me* to follow (I like to eat but not to cook), and the roasts are delicious!

"THYME FOR GOAT" ROAST

2–3 pounds boneless leg roast
2–4 garlic cloves, sliced
1 tablespoon rosemary
1 tablespoon thyme (fresh herbs are wonderful in this recipe)
a clay pot, covered baking dish, or oven baking bag

Insert knife into roast to create small pockets for the garlic slices. Add garlic to taste. Place roast and herbs in bag/dish/pot. Add 1½ cups water (can substitute white wine). Optional: add small peeled whole onions, small whole unpeeled potatoes, and/or carrots sliced in 6-inch chunks.

Roast at 325°F for approximately 1½–2 hours, or until internal temperature reaches medium rare (145°F) or medium (160°F).

Remove from oven, let rest, slice thin, and serve with cooking juices spooned over the meat. Leftover juices and meat make a wonderful soup or stew. Serves 4–6.

"THYME FOR GOAT" CURRY

1 teaspoon black mustard seed
1 teaspoon salt
2 teaspoons ground cumin
1 28-ounce can of chopped tomatoes (or 4 fresh tomatoes, chopped)
1 teaspoon fresh ginger (grated)
2 carrots
2 teaspoons ground coriander
2 stalks celery (optional)
1 teaspoon turmeric
1 cup fresh or frozen peas
1–2 garlic cloves (chopped/smashed)
½ cup cashews (unsalted)
1 large onion
1 large red pepper
2 pounds bone-in shank cuts, or stew goat meat

Sauté first six ingredients in ¼ cup oil or ghee (clarified butter) in large heavy Dutch oven.

Add onions and sauté, stirring until they are softened. Add goat meat, stir to cover the meat with the spices, and sauté for about 5–7 minutes. Add tomatoes, carrots, peppers, and celery. Add water just to cover, cover with a lid, then simmer for 1 hour, stirring occasionally and adding water if necessary.

Add the cashews and fresh or frozen peas, then simmer uncovered until sauce is thickened to your desired consistency. Flavor to taste with salt.

Serve with rice, flatbreads, or couscous.

CROCK POT GOAT CHILI

2 pounds ground goat
1 cup chopped onion
4 large garlic cloves, chopped
½ cup green peppers, diced
2 teaspoons dried cumin
½ dried turmeric
1 teaspoon chili powder
½ teaspoon salt
1 teaspoon ground black pepper
2 16-ounce cans of kidney beans (not drained) or leftover New England baked beans

In a large skillet, brown ground goat and drain. Add meat and all other ingredients to slow cooker. Cover and cook on low for 2 hours. Correct seasoning, then serve with coleslaw and cornbread.

STIR-FRIED CHEVON WITH GREEN ONIONS

⅔ pound *chevon* loin or leg, cut into thin slices
2 tablespoons sesame or safflower oil
12 green onions, cut into 1-inch lengths

Marinade No. 1
½ teaspoon garlic powder
2 tablespoons soy sauce
½ tablespoon sugar
2 tablespoons rice wine
2 tablespoons cornstarch

Marinade No. 2
3 tablespoons soy sauce
½ teaspoon sugar
½ teaspoon black pepper
4 tablespoons water

Cut meat into uniform ⅛-inch slices, 1½–2 inches long. Place meat in a sealable plastic bag or bowl with a leak-proof lid. Add marinade No. 1, then shake to coat thoroughly. Refrigerate at least 1 hour, shaking at least once.

When ready to cook, stir fry meat in 2 tablespoons sesame or safflower oil, stirring often until done. Add marinade No. 2 and green onions. Continue to stir fry until thoroughly hot; serve over warm rice. Serves 5.

Looking for more meal ideas? Additional sources for goat recipes are listed in the Resources section.

CHAPTER 11

IT'S YOUR BUSINESS!

If you've turned to this chapter thinking you will be given all sorts of rules, you're going to be disappointed. This chapter isn't here to tell you what to do, but instead to stress that what you're doing with meat goats—whether your enterprise is for profit, to break even, or for pleasure—is *yours*.

To help you figure out your startup costs, I've included the worksheet below. As a person who's learned the value of flexibility, I strongly recommend using a pencil to fill it out. (Having an eraser nearby, maybe even on the pencil, will probably come in handy too!)

STARTUP COSTS:
What You *Have* to Have Before You Get Your Goats

Secure pen/pasture to hold the goats $ ________
Feed (see Chapter 5) $ ________
Water (trough, hoses, frost-free on-demand waterer) $ ________
Minerals (feeders, the mineral itself, see Chapter 5) $ ________
Shelter (as necessary, see Chapter 4) $ ________
Safety from predators (see Chapter 7) $ ________
Subtotal of "needs" $ ________

It looks like both the young lady and the goat kid will be able to evade the paparazzi's cameras when they venture out in public! *Bryce Gromley for Gromley Farms*

Meat goats easily and happily traverse steep hillsides, devouring the seedpods of undesired vegetation. When the plants can't spread their seeds anymore, new weeds don't grow. *Weed Goats 2000*

THE MEAT GOAT HANDBOOK KEYS TO SUCCESS

- Be in agreement with your loved ones about your vision.
- Be independent in your thoughts and goals. Don't let anybody else tell you what to do—it's not their enterprise. Thoughtful counsel can be helpful, but inflexible pronouncements usually aren't.
- Be patient and true to your dreams. Remember, things that happen instantly aren't as likely to stick around (keep going and growing) as those that develop more slowly.
- First answer the "why" (your goals), then deal with the "how" (how to achieve them).
- If you're planning to sell products, first decide who the customers are and what you want to sell them. If you already know exactly what you want to sell, take the time to step back and fill in the blanks so that you don't miss something important.
- Whether it's for fun or profit, treat the enterprise as a business. Know what you're spending and what you're earning, so that you can avoid unpleasant surprises. And if you're forewarned about good news, think of all the fun you can have planning for its arrival.
- Spend money on moneymakers and goal achievers, because you will see them helping your progress toward your goals.
- Start with adequate money, energy, and time. Don't give in to the temptation to spend too much (or more than you have) of any of them. This seems so basic that most people agree with it and move on, but it's the sneaky trap that has caught many producers. Getting too big too fast is very difficult to overcome.
- Spend time with successful and happy meat goat producers. Their positive attitude will rub off on you.
- Fuel growth with your own money, energy, and time. It's all too easy to spend someone else's money, and most "someone elses" charge interest.

STARTUP COSTS:

What You *Want* to Have Before or Soon After You Get Your Goats

Kidding kit and toolkit	$ ________
Handling facilities	$ ________
Scale	$ ________
Subtotal of "wants"	$ ________

STARTUP COSTS:

What You *Wish* to Have Someday (can move up on your list if necessary)

Trailer, topper, goat tote, dog crates	$ ________
Subtotal of wishes	$ ________
Subtotal, startup costs (excluding animals) (add 3 subtotals above)	$ ________

ANIMAL COSTS

Does	$ ________
Doe kids	$ ________
Bucks	$ ________
Subtotal, animals	$ ________
Total that you need to have to "start" add 2 subtotals)	$ ________

See that last total number? If that number is smaller than what you have in your wallet, piggy bank, or bank account, and you don't need any of the money for groceries, that's great, you can "goat for it"! If that last number is larger than your available funds, you have two options: You can lower your startup costs by buying fewer or less expensive items, or you can increase the amount

In the vast acreages of the West, a herder and her horse move thousands of meat goats to their next target forage for the goats' enthusiastic consumption. *Weed Goats 2000*

of money that you have available to spend. (The expression "beg, borrow, or steal" comes to mind, but let's disqualify stealing as a good way to start a business or a hobby.)

OTHER WAYS TO HAVE PARTNERSHIPS WITH MEAT GOATS

- *Weed control*—Using goats to harvest unwanted or nuisance plants is rapidly gaining in popularity. Meat goats can slow down or reverse the spread of invasive species without the use of herbicides or pesticides. Herds of fifteen hundred or more goats are frequently needed for prescriptive grazing on the vast acreages of the West, whereas smaller herds would be extremely effective in the Midwest, South, and East Coast. If you'd like to learn more about prescriptive grazing, *The Targeted Grazing Handbook*, listed in the Resources section, is a good source of information.
- *Pack animals*—Wethers (castrated male goats) can be trained as pack goats. They can carry approximately 25 percent of their body weight as they travel with people who wish to hike and explore the outdoors.

Goat's milk soap is also a popular specialty product.

ESCAPE GOAT TOURS

Escape Goat Tours (www.EscapeGoats.us), in Escalante, Utah, is a permitted outfitter and guide service for the Escalante Grand-Staircase National Monument and Glen Canyon National Recreation Area. Located less than an hour away from Bryce Canyon, Escape Goat Tours offers a variety of day and evening hikes, backpack tours, photo tours, and catered pack trips with its friendly porters, the pack goats. People from across the country and around the world have enjoyed hikes through the slot canyons, the arches, and the seldom-seen back country, and to view the centuries-old pictographs on the canyon walls.

Guides and owners Shawn and Shari Miller moved to Utah in 2004 with Shari's experience in showing Oberhasli dairy goats and Shawn's desire to no longer be a truck driver. They founded Escape Goat Tours and have been enjoying showing visitors the beauties of the Escalante area with their string of friendly porters "with great personalities" ever since.

EWE4IC ECOLOGICAL SERVICES

Ewe4ic Ecological Services (www.GoatsEatWeeds.com) uses goats in a controlled grazing environment to gradually and naturally remove weeds and return the land to a healthy, natural ecosystem. "Goats are self-propelled entities that are able to go where no machine can and few humans dare," explains owner and founder Lani Malmberg. "Goats work during all seasons, in fair weather as well as when it is windy, dark, snowy, raining, and on all major holidays."

Lani has degrees in environmental restoration, botany, and biology and a graduate degree in weed science. Hooked staff in hand, she wanders the meadows, hillsides, and waterways of the West, pitting fifteen hundred Cashmere goats against pockets of unwanted weeds that infest the landscape. Lani and others feel that since chemicals have been used against weeds for almost half a century and there are now more acres of weeds than there ever have been we need a better way to heal the land.

Lani's work is an inspirational example of successful pesticide- and herbicide-free weed management, fire fuel load abatement, reseeding, watershed management, and land restoration. Her goats can eat 3 percent (on a dry matter basis) of their body weight per day, and consume, with no adverse effects, hemlock, poison oak, all of the tamarisk species, blackberry bushes, blooming yellow star thistle, and different mustard species. The goats' hooves till and aerate the soil and trample in their own fertilizer. And since goats are browsers whose diet consists of approximately 70 percent nongrassy species, they don't compete with cattle for the grass.

To date, the list of plants that the goats have been hired to control includes: Canada thistle, cheat grass, common tansy, common mullein, Dalmatian toad flax, dandelions, downy brome, Indian tobacco, kudzu, larkspur, leafy spurge, loco weed, musk thistle, oxeye daisy, plumeless thistle, poison hemlock, purple loosetrife, Scotch thistle, all of the knapweeds, sweet clover, yellow star thistle, and yucca.

- *Food and soap products*—You can make goat's milk ice cream as well as *chevon* snack sticks or jerky. Goat's milk soap is also a popular specialty product.
- *Construction*—Have your bucks peel logs for log-home construction in the eleven months they're "on vacation."
- *Kidskin products*—You can make kidskin hides, tanned hair-on from a winter harvest when the goat is carrying cashmere, as baby play rugs. You could bring the phrase "handling with kidskin gloves" back into fashion!
- *Therapy goats*—Visit shut-ins or rest homes with your goat ambassadors (more or less grownup bottle baby goats).
- *Ecotourism*—Get paid to have people visit your farm or ranch to learn about what your

In order to get to the next grazing area, goats will go through shallow water if their herding dogs ask them to; however, they will even more happily use a bridge. *Carolina Noya, Flying D Cattle Co.*

These Angoras and meat goats are harvesting unwanted vegetation in a *recharge* in Arizona. A recharge is where treated and purified water is allowed to sink back through the earth to the aquifer, and vegetation steals water. Meat goats to the rescue! *Brad Payne, Arizona EcoGoats*

With the herder and the guardian dog following the goats, these Wyoming goats head toward winter headquarters.
Carolina Noya, Flying D Cattle Co.

Life doesn't get much better than this to a meat goat: good forage, trees for shade, and owners who care about them.
Carol DeLobbe, Bon Joli Farm

It's time for the kids to sleep and dream about tomorrow's fun and games. *Kim Hunter, Fossil Ridge Farm*

goats do for the ecosystem and the human food supply.

- *Culinary tourism*—Partner with a local chef (and a vintner, and cheesemaker, and ice cream maker, and . . .) to offer healthy and delicious goat meat, milk, and cheese products for immediate consumption and to take home from local specialty stores and restaurants.

Sometimes the list of possibilities for meat goats seems longer than the days are. Many of the opportunities on that list could not only be good business for you but examples of making mutually beneficial connections with others. For example, it's so easy for rural people and urban people to misunderstand each other, get defensive, and disqualify the other's right to feel or think the way that they do. Except that the two groups need each other, but can easily forget that. The urban customers buy the rural product, which brings revenue to the rural areas and food to the urban areas. Since meat goat people are such good people, it should be much easier for us to reach out to our customers and make progress for ourselves and for them.

KNOW YOUR GOAT VOCAB

The scientific name for *goat* also gives us many wonderful words. The Latin name for the domestic goat is *capra hircus* and the *capra* can be found at the root of "capricious," "capriole," and "caper," and "kid" (as in young goat), "kid gloves," and "kidding around."

Capricious—When something is capricious, it changes direction suddenly and without warning, sometimes for no apparent reason. Usually *capricious* is used in a somewhat insulting way, but when a goat does it, it's a survival mechanism. Goats run and bounce from side to side rather than running in a straight line because their zigzag path makes it much more difficult for a predator to successfully catch them. So there is, in fact, method to their madness!

Capriole—The capriole is the beautiful airs above the ground movement performed by the Lipizzaner stallions, the white dressage (equine ballet) performers. Very soon after your goats get home and one jumps off of anything more than a few inches high, you may well see the goat, in midflight, kick its back legs out behind itself. That is the capriole movement done by the Lipizzaner stallions, and having been privileged to see both horses and goats do it, I think the goats do it with more joy. And nobody taught them how to do it! I've never read or heard why goats kick out behind themselves when leaping off of something. Maybe they do it just because they can!

Caper—The dictionary says that *caper* is a shortening of *capriole*, and it means "to leap or skip about in a sprightly and playful manner; prance; frisk; gambol; a prank or trick; harebrained escapade." All the way up to "gambol," it could be describing goat kids at play, and I'll vote that *caper* was used to describe a harebrained escapade by people who were jealous of goat kids' spontaneity and *joie de vivre* (joy in living).

Kid gloves—Treating someone very well, or taking care to be extremely gentle with them, is frequently called "treating them with kid gloves." Since kid gloves were made from the skin of a kid (young goat), they were softer and finer than other gloves, probably more expensive than other gloves, and thus became a symbol of elegance in the upper classes in the early 1800s. Since kidskin is so much finer (thinner) than other leathers, it is not as strong, and the expression "to treat someone with kid gloves" might actually have started as a description not of how gently you were treating the person, but how gently you were treating the gloves around your hand!

To forage managers, the plants pictured here are invasive weeds. To meat goat managers, they are terrific goat food. *Jean-Marie Luginbuhl, North Carolina State University*

Black locust.

Kudzu.

Honey locust.

Multiflora roses.

Honeysuckle.

Sericia lespedeza.

RESOURCES

"HOW TO" INFORMATION BY CHAPTER

Chapter 5—Find information about using a Pearson Square at www.ext.colostate.edu/pubs/livestk/01618.html and prechel.net/formula/pearson.htm.

Chapter 6—Listings of upcoming FAMACHA training, to learn how to "eye-check" your goats and get your handy chart (as shown on page 114), can be found by visiting www.scsrpc.org/SCSRPC/Workshops/fworkshops.htm.

Chapter 7—*DogLog*, a newsletter from the Livestock Guardian Dog Association, details how livestock guardian dogs mature and work. For copies visit www.ProfitableMeatGoats.net or call 406-466-5952.

Chapter 7—*Dogs: A Startling New Understanding of Canine Origin, Behavior & Evolution,* by Raymond Coppinger and Lorna Coppinger. A "book that explains what dogs are, why they are different from wolves and from each other, and how their relationship with people can be enriched so both species can benefit."

Chapter 9—Learn how to use a Burdizzo by visiting www.ansi.cornell.edu/4H/meatgoats/castrating/castrate.htm.

Chapter 9—Dr. Rick Machen's fact sheet on creep feeding goats, published by the Texas A&M University system, is available online at www.motesclearcreekfarms.com/SABoerGoats/asp/4H/Creep-Feeding-Goat.pdf.

Chapter 9—Learn how to disbud by visiting www.HoeggerGoatSupply.com/info/disbud.shtml.

Chapter 9—Learn how healthy, happy goat kids behave by visiting www.GigglewiththeGoats.com (DVD available).

Chapter 10—Download the LSU Publication 2951 from www.LSUAgCenter.com.

Chapter 10—*The Complete Book of Butchering, Smoking, Curing, and Sausage Making*, by Philip Hasheider (Voyageur Press, 2010).

Chapter 10—Find delicious goat meat recipes at www.jackmauldin.com/goat_recipes.htm.

Chapter 10—Choose from several flavors of Laloos's 100 percent goat's milk ice cream at www.laloos.com/getsome.php.

PERIODICALS

Goat Rancher (www.GoatRancher.com, 888-562-9529). The magazine of America's commercial meat goat industry, and the only magazine published by goat ranchers, for goat ranchers.

The Stockman Grass Farmer (601-853-1861 or www.StockmanGrassFarmer.com). Serves as an information network for grassland farmers sharing the latest in intensive-grazing technology and pasture management.

BOOKS

A Compilation of the Wit and Wisdom of "The Goat Man" by Dr. Frank Pinkerton, available through *Goat Rancher* magazine (www.GoatRancher.com or 888-562-9529).

Profitable Meat Goats Conference reference book, DVD set, and printed transcriptions available. Purchase from www.ProfitableMeatGoats.net or by mail at 1380 Hwy 220, Choteau, MT 59422.

Natural Goat Care, by Pat Coleby (Acres U.S.A., 2001). Includes eco-farming techniques for commercial-scale farming.

ONLINE RESOURCES

The Spanish Goat Association (www.SpanishGoats.org or 540-687-8871). The Spanish Goat Association welcomes all those interested in Spanish landrace goats.

The American Livestock Breeds Conservancy (www.albc-usa.org or 919-544-0022) is a nonprofit membership organization working to protect over 180 breeds of livestock and poultry from extinction. Included are asses, cattle, goats, horses, sheep, pigs, rabbits, chickens, ducks, geese, and turkeys.

Premier (www.premier1supplies.com or 800-282-6631) offers farm supplies, including electric fence and netting, ear tags, and other sheep and goat equipment.

Langston University (www.luresext.edu/goats/index.htm) produces the *Meat Goat Production Handbook* and a pocket card for body condition scoring. The school also holds a Goat Field Day, an annual informational and educational event in Langston, Oklahoma.

Breed registries:

Boer—includes the USBGA (U.S. Boer Goat Association), IBGA (International Boer Goat Association), and ABGA (American Boer Goat Association).

Kiko—includes the AKGA (American Kiko Goat Association), IKGA (International Kiko Goat Association), and NKR (National Kiko Registry).

Sheep and Goat (www.sheepandgoat.com). Susan Schoenian, University of Maryland Extension sheep and goat specialist, has created an extensive list of links covering all aspects of sheep and goat production.

Targeted Grazing Handbook (comes with DVD). Available from the American Sheep Industry (www.sheepusa.org/Online_Education).

ATTRA (formerly known as Appropriate Technology Transfer for Rural Areas, found at attra.ncat.org). The National Sustainable Agriculture Information Service is your source for the latest news in sustainable agriculture and organic farming news.

Holistic Management International (HMI) (www.holisticmanagement.org or 505-842-5252) provides, promotes, and teaches holistic land management, which works in concert with natural processes. HMI has helped restore damaged grasslands to health and sustainability and increase the productivity and profitability of ranches and farms.

GLOSSARY

Abattoir: Facility that processes live animals into carcasses or cuts of meat. Also called a harvest facility, processing facility, or kill facility.

Abomasum: Fourth or true stomach, where hydrochloric acid digests food particles that have been repeatedly chewed (cudded) by a ruminant.

ADF: Acid detergent fiber, the least digestible parts of the plant cell wall.

ALBC: American Livestock Breeds Conservancy

Anemia (anaemia): A deficiency of red blood cells in hemoglobin; means blood cannot carry oxygen to muscles and organs.

Anestrus: When a doe does not come into heat.

Animal unit: One thousands pounds of animal body weight.

Animal unit month (AUM): How much forage an animal unit needs to consume in a month.

Backgrounding: Preparing animals for optimal condition or feed conversion (typically prior to entry into a feedlot).

Backstrapping: Feeling a goat's short ribs to determine the amount of flesh it's carrying.

Banding: Castration by putting a small elastic band around the skin between the testes and the body.

Belling: Putting a collar with a bell on a goat for locating the animal by sound.

Billy: See *Buck.*

Bloat: The buildup of gas in a goat's rumen.

Body condition scoring: A numerical scale for evaluating the physical status of animals.

Breech: Being born back legs first.

Broken mouth: A goat that is losing teeth due to old age.

Buck: A male goat that is able to produce offspring.

Buck effect: When female goats are triggered to go into heat due to the scent of a buck in rut.

Buckling: A young male goat.

Burdizzo: A tool used to sever (by clamping) the cords that take blood to the testes in male goats.

Cabrito: Goat meat, usually from a young goat.

Caprine arthritic encephalitis (CAE): A viral infection of goats that may lead to chronic disease of the joints and on rare occasions encephalitis (inflammation if the brain).

Carrying capacity: How many pounds of animals a given area of land can support.

Caseous lymphadenitis (CL): A contagious goat disease that causes abscesses, often in the throat area. Also known as cheesy gland.

Cattle panels: Sixteen-foot-long welded wire panels, fifty-two inches tall with ten horizontal rods. Not completely effective at holding small goat kids.

Chelated: Minerals that are protected from blocking agents so that meat goats can metabolize them.

Chevon: Name for goat meat, based on the French *chevre* (goat).

Chilled weight: Weight of an animal carcass following cooling.

Coccidiosis: Disease caused by excessive quantities of oocytes, a protozoa in the goat's digestive tract. Typically caused by eating from contaminated ground or stress. Can cause diarrhea and death.

Colostrum: First milk, rich in antibodies and minerals that newborn kids need to survive and thrive.

Combination (combo) panels: Sixteen-foot-long welded wire panels, fifty-two inches tall with thirteen horizontal rods. Very effective at holding small goat kids.

Concentrate(s): Feeds other than forage, such as grains.

Conjugated linoleic acid (CLA): A newly discovered good fat that may be a potent cancer fighter.

Cost per mile (fencing): How much it costs to put up a mile of fence.

Crayon: A colored wax block that, when placed in a buck's marking harness, makes a mark on the haunches of any doe bred by that buck.

Creep feeding: Setting up an enclosure that allows small (typically young) goats to creep in and get extra feed while keeping larger goats out.

Cryptorchid: Male animals with one testicle in normal position and one retained up in the abdominal wall. Thought to be hereditary.

Cudding: When a ruminant animal regurgitates, rechews, and reswallows its food.

Culling: Removing certain animals from the herd that are not deemed as worthy as others.

Cystic ovaries: Ovaries that are incapable of producing eggs.

Deep cycle battery: Marine battery that can be used to power electric fences.

Diatomaceous earth: A soft sedimentary rock that is easily crumbled into a fine white to off-white powder; reputed to reduce internal and external parasites.

Disbudding: Preventing an animal's horns from growing.

Doe: Female goat that is able to produce offspring.

Doeling: Young female goat.

Drenching: Giving a goat something orally, typically a medicine or dewormer.

Dry matter basis: The contents of feed with all moisture removed.

Eight-tooth: A goat whose teeth (eight total) are all mature. Typically approximately four years old.

Electrolytes: A powder that you mix and allow the goat to drink; offers the animal beneficial minerals and microorganisms, and stimulates energy levels. Electrolytes should not be fed to kids within four hours of nursing.

Ennobled: An animal that has passed a visual inspection and attained a requisite number of points through competitive (show) placements.

Enterotoxemia (overeating disease): Infection caused by toxins released in the intestinal tract due to a proliferation of clostridium perfringens type C & D bacteria. The bacteria are normally present in small quantities, but when an animal gorges itself on rich, unfamliar foods, C & D can flourish.

Estrus (oestrus): Period of sexual fertility and receptivity. Also known as being in heat.

Expose: Allowing a female goat to be bred.

Eye check: Looking at the color of a goat's inner eyelid to check for anemia.

Famacha: A visual evaluation system that rates a goat's level of anemia by the color of its inner eyelid.

Fishtail teat: A teat with two orifices. Frequently causes nursing difficulties for kids.

Flehmen response: Behavior of a breeding male wherein the upper lip is curled back even closer to the nasal passages than usual.

Floppy kid: Rumen acidosis. Causes loss of muscle control and a "floppy" appearance. It is thought that floppy kid is brought on by lack of exercise or constant access to the mother's milk, which causes the kid to get a new meal of milk before the previous meal is digested.

Flushing: The practice of feeding does extra rations in the weeks or month before breeding. This prompts the doe to ovulate more prolifically, due to weight gain.

Formalin: A 10–12 percent formaldehyde solution used in cleaning CL abscesses.

Four-tooth: A goat that has two sets (four total) of mature teeth, typically about two years old.

Frothy bloat: The buildup of frothy gas in the rumen, typically caused by the consumption of alfalfa or clover in warm, wet spring conditions.

Generation interval: Time between the birth of one animal and the birth of its offspring.

Gestation: The period of time offspring is carried in the womb.

Halal: Satisfying the requirements of Muslim law.

Hanging weight: The weight of an animal carcass immediately after processing and inspection.

Heat cycle: Length of time between periods of estrus.

Heterozygous: Having different genetic traits.

High tensile: Electric fencing in which the wires are tensioned (and thus a more effective physical barrier).

Hock: The knee in an animal's back leg.

Hog panels: Sixteen-foot-long welded wire panels, thirty-four inches tall with eleven horizontal rods.

Holistic resource management (HRM): The philosophy that all resources are part of one system and are interrelated and interdependent.

Homeopathic: A health-care system in which a patient is given minute doses of natural substances that in larger doses would produce symptoms of the disease itself.

Homozygous: Having the same genetic traits.

Horn blocker: Typically a 12-inch length of stick or PVC attached with duct tape to the base of a goat's horns to prevent the goat from putting its head through a fence and getting stuck.

Hornset: How a goat's horns are shaped and positioned on the head.

Hot shot: Electric prod used to direct animals.

Hot weight: The weight of an animal carcass immediately after processing and inspection.

Hot wire: Filament of electric fencing.

Humane: An action or behavior based on kindness rather than cruelty.

Kidding: Goats giving birth.

Kosher: Satisfying the requirements of Jewish law.

Lamb paint: Food-grade spray paint that can be used to mark an animal; typically remains visible for weeks.

Listeriosis: A bacterial disease that causes weakness on one side, frequently making the animal walk in circles. Often fatal, occasionally an infected animal can be saved with massive doses of antibiotics.

Liveweight: The weight of a live animal.

Low impedance: A type of electric fence charger that produces a high-amperage pulse and can overcome some fencing shorts.

Low tensile: Electric fence wire that is not under much tension. Unlike high-tensile fencing, it offers very little physical resistance, but because the fence is electrified or "hot," it is very effective in containing animals that are familiar with it.

Maintenance: Keeping the same body weight, neither gaining or losing weight.

Marine battery: See *Deep cycle battery.*

Marking harness: A harness that can be strapped around the chest of a breeding male; typically includes a crayon that makes a colored mark on the haunches of any female the male mates with.

Meconium: The tarry, sticky blackish first feces (poop) passed by a kid.

Milk down: To lose weight because of milk production.

Milk tooth: (Description of age) When all eight teeth are of equally small size, typically in goats under a year old.

Minerals: Inorganic compounds that meat goats require for survival and good health.

Myotonic: A breed of meat goat whose muscles temporarily lock up when the goat is startled. Also known as fainting goats.

Nanny: See *Doe.*

Naturally polled: A goat that will not grow horns, sometimes called a muley. There is a sex-linked trait for hermaphroditism associated with naturally polled goats.

Offset: An electrified wire held away from a fence (typically a physical fence).

Omasum: The third stomach of a goat. Removes water from food traveling through the digestive system.

On the rail: The term for an animal carcass hanging and ready to be (or having been) weighed.

Parturition: Giving birth.

Pearson square: Food ration formulation procedure.

Pinkeye: Infectious keratoconjunctivitis. Causes goats' eyes to run and become cloudy or whitish to reddish.

Predator pressure: How much effect predators have on any given group of livestock.

Premise ID: The numerical code assigned to each goat-producing location in the American scrapie eradication program.

Presentation (at birth): How a kid is physically positioned at birth.

Reproductive efficiency: The pounds of offspring born and raised per pound of mother.

Reticulum: The second stomach of a goat, functions as a sieve.

Rumen: The first stomach of a goat, which operates as a fermentation vat and begins the digestive process.

Ruminant: An animal that chews its cud.

Scrapie: A fatal degenerative disease of sheep and goats related to bovine spongiform encephalopathy (BSE or mad cow disease) and chronic wasting disease of deer.

Scrapie number: Premise ID number that U.S. meat goat producers are being asked to include on their goats' ear tags to assist in the eradication of scrapie.

Six-tooth: A goat that has three sets (six total) of mature teeth, typically about three years old.

Soremouth: Contagious ecthyma (called *orf* in sheep), a viral disease in the pox family that is highly contagious both to goats and humans. A goat is immune to the disease after having had it.

Standing heat: When a doe is not only in heat, but is also willing to be bred by a male goat.

Stargazing: When a goat holds its head up abnormally (as if looking at the sky). Behavior can be caused by a deficiency of thiamine (B vitamin).

Stocker (kids, does): Market goats purchased for additional feeding, typically in a pasture.

Tannins: Chemicals with astringent qualities, produced by certain plants. Used in the tanning of leather, tannins show great potential for reducing parasites in goats.

Teaser buck: A vasectomized male goat that smells and acts like a breeding male. Can be used to trigger heat cycles in does.

Total digestible nutrients (TDN): The energy value of feedstuffs.

Toxoid: An inactive toxin given as a vaccination to stimulate production of antibodies.

Two-tooth: A goat that has one set (two total) of mature teeth, typically about one year old.

Wattles: Small, hairy, frequently bell-shaped appendages of flesh usually hanging below the throat of a goat. Their function is not understood.

Wax plugs: Found in the end of a pregnant doe's teats in the days before birth. They keep the udders free from foreign bodies and prevent milk leakage.

Wether: A neutered male goat.

Winter stasis: When an animal temporarily stops growing during the coldest winter months.

Withdrawal: The time needed to ensure that medications (including dewormers) have been completely metabolized by an animal's body, and thus can no longer contaminate its meat.

ACKNOWLEDGMENTS

Simply in the order that I have been privileged to be allowed to learn from them, Mam and Pap, for encouraging my explorations, secure in the knowledge that they were always there (and that they would fish me back out of the lion's pen at the zoo). Big sister Anna, for showing me how to access the worlds that were beyond the doorway of the printed word. Doris, Kim. The 4-H program for the years with Seeing Eye puppies. Erve Meenderinkboer, Marilyn and Francois for the magnificence of equitation. Kyra, for ladyship lessons as well as dressage lessons. Lieutenant Commander Bob and Natasha for the presentation to the amazing livestock guardian dogs. The dogs for opening the gateway to the goats. Dr. Ray and Lorna, for patiently helping me separate the lore from the reality. Bron, for the clarity of the view of the meat goat world. Pat, for the steadfast friendship and unruffled counsel. Dr. Barbara, for helping me learn how to learn from the goats. Allan, Suzanne, and the whole Stockman Grass Farmer organization, who teach about the way we want to operate. Tom and Meta for helping me see the beauty of the big picture. Dr. Frank, I was too listening! Terry and Mary at the Goat Rancher, for the gentle and benevolent direction. Gamal, for proving that good people can be found everywhere and that a humane end for our animals is the correct next step in their journey. Leslie, thank you, and also for the Spanish Goat Association. Priscilla, thank you for the laughter and the reminder that perseverance is the right choice. Vaughn, thank you for reminding me that flights of fantasy are good, but it's best to land back on solid, well-tended ground. Cori, for the lessons in how to travel well. Sara and the kids, for caring for the kids. Tanner and Faithe: if I manage to grow up, I want to be like you. Dr. Jason, thank you for the amendments to the correctness. Joe and Jonathan, you're really good teachers. Jon, Brent, and Carrie, thank you. Dr. Steve for setting such a good example as well as giving all the information about meat goats. To all of those who shared pictures, thank you, thank you. Danielle, Anitra, and Melinda, it's not only easy to say that *The Meat Goat Handbook* wouldn't exist without your direction and assistance, it's absolutely true. Readers, thank you for needing the book and letting me discover how much there still is to learn while putting it together for you. To my best friend, the love of my life and my husband: thank you for our teamwork.

APPENDICES

SHOWING: GOOD POINTS FOR AND GOOD POINTS ABOUT

How to Make a Good Show

Have the goat properly trained, conditioned, and fitted before the show.

Dress appropriately with safe, comfortable shoes.

Always be courteous to the judge, fellow exhibitors, and show staff.

Never stop showing the goat, even before you enter the ring: Remember that the judge's first impression of your animal is when you get to the gate. Stop briefly if possible and have the goat standing alert and square, then enter when the judge indicates he or she is ready.

Hold the lead close to the collar and keep the collar high on the goat's neck to maintain optimum control and encourage the goat to walk with its head up. Lead the goat at a slow, relaxed pace to allow the judge to analyze the conformation and movement.

Keep your goat spaced from other goats in the ring as much as possible.

Be calm and confident. Acknowledge the judge with eye contact and a smile, and be sure to follow the ring steward's instructions.

Make sure that your goat is stood up at all times (which is square with its head high and its body in a straight line, with its side to the judge until instructed to have the goat face the judge or face away from the judge).

Always keep the goat between you and the judge, and if you need to change sides, move around the goat's head. Never walk around behind the goat.

Have fun in the ring and always be a good sport whether you win or lose.

Why Many People Show

Showing is a fun social activity, and they enjoy the fellowship and friendly competition.

Showing increases confidence and teaches good sportsmanship.

Sometimes they show goats that are just fun to show. It's not always about winning; sometimes it is just to have fun, as some goats just love being in the ring.

Showing is an excellent way to get outside opinions about the conformation of your goats and information about improving your breeding program.

Doing well in the show ring helps to promote your herd.

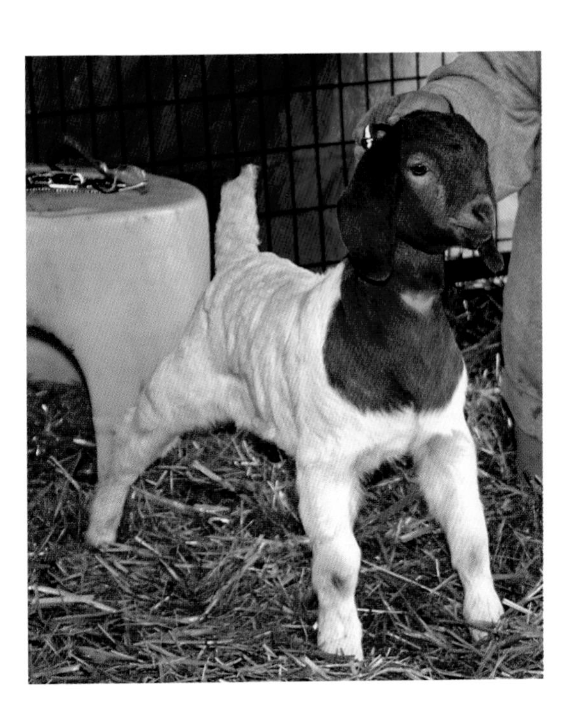

MEAT GOATS IN SOUTH AFRICA

Many people are surprised to know that South Africa has significantly more goats per person than the United States, despite the countries' vast difference in size. The chart below gives more details.

How did all those goats in South Africa develop? Were there native South African goats? And who made modern South African goats? Looking only at recent history (the past few centuries), different groups across the African continent raised different types of goats. Selective breeding was used in the development of certain breeds, such as the Boer goat. *Boer* is the Afrikaans word for farmer; Afrikaans comes from Dutch, which was spoken by the settlers who came to South Africa and Namibia in the seventeenth century.

Just as many meat goat producers in the U.S. prize the Spanish goat, some producers in South Africa are now working to protect the remaining Skilder goats (skilder is Afrikaans for "painted"). An article by

	U.S.	South Africa
Square miles	3,537,441	471,443
Population of people	300 million	49 million
Population of goats	2 million	5 million
Goats per square mile	0.57	10.6
Goats per 10 people	0.07	1.0

Black and white Skilder goats. *Natural African Indigenous Veld Goats*

Dr. Quentin P. Campbell in the No. 9, 1990, issue of South Africa's Boerbok Nuus tells how Skilder goats were the raw material from which the Boer, Savanna, and Kalahari Red goats were developed. Now there is an organization of South African producers who are working "to promote the pure preservation of the eco-types," which is how they classify different types of goats that are found living and reproducing successfully in different areas of the South African veld, or open grasslands.

Northern Cape speckled goats. *Natural African Indigenous Veld Goats*

A Northern Cape speckled buck. *Natural African Indigenous Veld Goats*

Nguni (Mbuzi) goats. *Natural African Indigenous Veld Goats*

Kunene goats. *Natural African Indigenous Veld Goats*

INDEX

ABOUT THE AUTHOR

Yvonne Zweede-Tucker has been enjoying raising meat goats for twenty years since founding Smoke Ridge. Her writing has appeared in *Goat Rancher* magazine and the *Stockman Grass Farmer*, and she organized and was one of the presenters at the Profitable Meat Goats conference in Indianapolis. Yvonne lives with her husband, Craig, on their goat ranch in Choteau, Montana, and is sure that "the outside of a goat is good for the inside of a person." www.smokeridge.net.